INSTRUCTOR'S MANUAL
WITH SELECTED
SOLUTIONS FOR

APPLIED
COMBINATORICS

SECOND EDITION

ALAN TUCKER
STATE UNIVERSITY OF NEW YORK
STONY BROOK

JOHN WILEY & SONS
NEW YORK / CHICHESTER
BRISBANE / TORONTO / SINGAPORE

CONTENTS

PREFACE TO SECOND EDITION

The major change in the second edition of this text is the reversal of the two main parts, enumeration and graph theory. Until now, all combinatorics texts (that contained both enumeration and graph theory) have put the graph theory second. However, the MAA Panel on a General Mathematical Sciences Program suggested a model curriculum for Discrete Methods courses that puts the graph theory first. This writer was on that panel and was convinced by others of the merits of reversing the traditional order. Before giving the reasons for the change, it should be noted that IT IS STILL EASY TO TEACH THE ENUMERATION PART FIRST BY STARTING IN PART TWO.

a.) The hardest counting problems come right at the beginning of the enumeration part (the ad hoc problems in the middle of Chapter V), while the graph theory begins more easily and generally is less difficult. Students will learn the counting material better if they have some previous experience with combinatorial reasoning and case-by-case arguments, gained from doing the graph theory first. Two years ago, this writer started teaching the graph theory first and found that students did do better in the initial counting problems.

b.) There are many nice problems that combine enumeration with graph theory, e.g., counting the number of colorings of a graph (by ad hoc methods and by inclusion-exclusion). In these problems it is the enumeration that is the difficult part, while the graph theory component is usually little more than some definitions. These combined problems are really counting problems and belong in counting chapters. Then the necessary graph theory needs to be introduced previously.

There is one caveat that might persuade some instructors to keep the old order. If most students in the course are concurrently taking computer science courses that have major programming projects due at the end of these courses, then it may be desirable to have the more difficult part of the combinatorics course come in the first half-- when the students will have adequate time for the homework exercises.

Instructors using this text are encouraged to write the author (% Department of Applied Math, SUNY, Stony Brook NY 11794) with their reactions to the reorganization in the Second Edition and any other criticism or comment.

Part I   INTRODUCTION

1. Preparing to teach an applied combinatorics course for the first time.

The good news first.   There is little new material an instructor need learn to teach an undergraduate applied combinatorics course. Most chapters in this book are built around a few simple formulas or principles. This is a problem-solving text, not a theory text. Advance preparation consists primarily of working out for oneself problems in the book. The principal attribute required of an instructor for this course is a disciplined, logical mind.

Now the bad news. There is much that occurs in the classroom in this course for which an instructor cannot prepare. The critical tasks are: a) determining the reason underlying a student's wrong answer or misconception about a counting principle; and b) being able to spot alternate forms of an answer--expressions which look very different from the standard answer but still are correct. Some exercises have 4 or 5 equally correct analyses yielding different looking expressions. The combination of "debugging" mistakes and unexpected forms of an answer also may arise, when a student uses an original analysis but makes mistakes along the way.

In this course, the job of the instructor is not so much to present a right analysis of problems as to help students learn how to devise a right analysis for themselves. This means, answering student questions (especially about homework problems) should be a major part of the class time. Trying to answer all these questions promptly in class is clearly frought with danger. The best way, time permitting, to answer a "why can't you do it this way" question is for the instructor to work slowly through the problem, starting from first principles and using a small concrete example, and follow the suggested line of attack until a fault is found or a valid, correct answer is obtained. The instructor should encourage other students to participate in the analysis. Sometimes the instructor may get stuck and should quickly say that he/she needs time to think about the question in order to give a clearer answer. Don't try to hide the fact that the material can be tricky for the instructor, too.

One of the best ways to prepare for unforeseen questions about homework is to try to solve exercises in advance many different ways. An even better tactic is to encourage students to come to the instructor's office to ask some of their questions about homework before it is discussed in class. The same questions inevitably come up again in class, and now the instructor is all set with pretested, clear answers.

2. General teaching suggestions

The purpose of this Instructor's Guide is to help insure a successful experience for an instructor teaching this material. An applied combinatorics course using this text is ideally suited to a style of "interactive" mathematics teaching that gives students a more active and, hopefully, more rewarding role in the classroom. The following principles underlie this style of teaching.

First, introductory courses in any mathematical subject, especially courses serving many non-math majors, should try to leave students with a favorable attitude toward the subject and the learning experience. No matter how important the mathematical methods, students often avoid using them, or learning more about them, if they disliked the way they were taught.

Second, all applied mathematics courses should motivate topics with applications and develop concepts intuitively at first. In a first undergraduate course in an applied subject, a problem-solving approach is desirable (similar to the standard problem-solving approach in calculus). An instructor's enthusiasm for an elegant theorem may influence some students to study the mathematics for its own sake, but first of all, an instructor's outlook in class should reflect the needs and point of view of the majority of students, typically a problem-solving point of view.

Third, instructors should seek to develop good rapport with their students. They should foster student interest in the subject with their enthusiasm for teaching the material and with their encouraging answers to students' questions (even silly questions). If students believe that the instructor is interested in their ideas and questions, they are likely to give the instructor constructive feedback that will make the course even better.

The success of the Stony Brook applied combinatorics course, which last year was the largest upper-division math course (pure or applied) at the school, is due in large measure to the interactive teaching used in the course, as well as the course's applied, problem-solving point of view.

3. Goals of this text

This text introduces students to combinatorial problem-solving. The most important technique for a combinatorial problem-solver is simply disciplined, logical analysis of "word" problems. Such reasoning is the foundation for building simple mathematical models of problems: models implicit in counting expressions built of sums and products of binomial coefficients, in generating funtion models, in recurrence relation models, or in graph models. The way for a student to learn this logical thinking is by working many, many exericses. Students will use this reasoning often, consciously and unconsciously, in computer science, in operations research, and in probability and statistics. A logical mind will serve a person well in any field. To some, such a goal makes this text little more than an IQ enrichment book. Whatever the interpretation, <u>the goal of developing such logical reasoning in a problem-solving framework should be foremost in the instructor's mind at all times.</u>

A secondary goal of this text is to introduce students to the basic concepts and tools of enumeration and graph theory: counting methods such as generating functions, recurrence relations, inclusion-exclusion formula, and Polya's enumeration formula; and basic types of graphs, such as planar graphs and trees, and the uses of graphs in computer science and operations research. Several of these topics have special pedagogical value. For example, recurrence relations develop recursive reasoning (so important in computer science). The inclusion-exclusion formula is a grand exercise in applied set theory. Polya's formula is a practical (and fairly painless) introduction to group theory.

Finally this text seeks to achieve these learning objectives in as appealing a manner as possible, with card problems, logical puzzles, and the game of Mastermind. This writer's experiences in teaching applied combinatorics consistently re-affirm the importance of giving problems interesting settings.

4. Student Background

Many students find the transition from calculus courses to a combinatorics course very frustrating. They have become conditioned to technical "plug in" problems, applying a given integration technique to various functions. They are not accustomed to having to figure out how a problem should be solved. A combinatorics instructor often hears the complaint from students that the textbook is unfair because the homework exercises are so different from the worked—out examples--they expect that the homework problems should be solvable by mimicking calculations in the examples. Related difficulties may arise for math majors used to theorem-proving courses. They may have trouble developing their own simple models for analyzing word problems.

Conversely, there are some students for whom combinatorial problem-solving comes very easily. However, as in most other math courses, students who coast along will see their grades deteriorate when other students catch and surpass them by dint of hard work.

Problem-solving is a skill that virtually any student can learn with enough effort (as opposed to pure mathematics, such as modern algebra, whose proofs never make sense to some students).

5. Note-Taking

A correct solution to most combinatorics problems is often "obvious" once seen. As a consequence, students often copy down little or no explanation of the reasoning behind an answer. Consider the problem, how many different non-empty subsets (of any size) are there of the integers $1,2,\ldots,n$. The answer is 2^n-1, which can be explained: any subset (including the empty set) can be represented as an n-digit binary sequence with the i-th digit = 1 if and only if i is in the subset; and there are 2^n n-digit binary sequences and hence 2^n subsets, 2^n-1 of which are non-empty. If a student understands the binary sequence model quickly, then the 2^n-1 answer will be obvious and the student may write down only the numerical answer with no explanation, or possibly the phase "model as binary sequence." But several days later when students read over their notes, the expression 2^n-1 may now mean nothing, nor may the phrase "model as binary sequence."

The instructor at first must remind students to be careful to take good notes, and the instructor should pause at times to allow students the chance to write out explanations.

When the answer to a problem is a product or sum of binomial coefficients, etc., it is helpful to annotate the answer as follows. For the problem, find the probability that a random 5-card hand is a full house (a triple and a pair), one might annotate the answer thus:

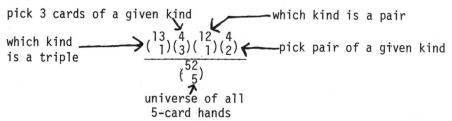

6. Homework Assignments and Tests

In this course, the problems are the message. Exams are meant primarily to measure a student's proficiency at solving problems. A natural consequence of the previous statement would seem to be that homework should directly make up a major portion of the course grade. However, so much of the course centers around the homework and how a student learns from homework mistakes. If for each section, there were a homework assignment, class discussion of the assignment, and finally a second assignment on the material, then counting the homework grade on the second assignment would be reasonable. This writer often creates a de facto second assignment by going over half of the exercises in an assignment in one class and then letting students have extra time to go back and rework, if necessary, the other exercises and hand in the whole assignment in the next class.

The most important aspect of discussing students' homework solutions is the <u>wrong answers</u>. The old adage "learn from one's mistakes" is the essence of modeling. Mathematicians do no divine right answers. They have ideas and sort through them in an evolutionary way until all the steps needed for an answer are appropriately assembled. Along the way, many wrong ideas must be discarded. Typically, it is by finding errors in an initial guess that one gains the insight that leads to the correct analysis. Students learn only so much from seeing right answers. The instructor should help students learn to find right answers from initial wrong answers. (Asking students to offer forth their wrong answers for critical, albeit constructive, analysis is a tricky business that requires good student-teacher rapport).

The main difficulty with tests in this course is that the homework problems often require a lot of time, a limited commodity at test time. A secondary problem is that after doing some graph theory, students quickly forget how to analyze counting problems. Thus it is hard to give a cumulative final exam. This second problem has forced many (but not all) combinatorics instructors to have non-cumulative last tests. The first problem can sometimes be solved by scheduling a 2-, or 3-, hour special time for a mid-term test. If this is not possible, two one-hour mid-term tests plus a longer last test (during final exam week) are the best one can do.

Reviewing for tests is an important learning period in the course. The week before a test, students are eager to solve lots and lots of practive problems. Students discover that they now can solve problems that tricked them earlier. It is helpful to hand out a copy of last year's test (or a realistic sample test).

An extensive set of sample homework assignments and tests is given in Part Three of this Instructor's Manual.

PART II COMMENTARY ON TEXT

Chapter I Elements of Graph Theory

This chapter introduces students to graphs. A relatively informal approach is used, which avoids the litany of definitions and theorems found in most introductions to graph theory. Section 1.1 illustrates the diverse uses of graph models. Section 1.2 uses the question of when are two graphs isomorphic to explore the intrinsic structure of graphs. Section 1.3 presents an important simple theorem about graphs. Section 1.4 examines a natural graph property, planarity.

This is the most important chapter about graphs, and all its sections should be covered. Although graphs are assuming an ever-growing role in computer science and have always been important in operations research, most students will never have seen graphs before. The whole concept of a non-numerical mathematical model will be new to many. The examples and exercises are generally easier than the

5

problems in later chapters. This chapter's problems aim to illustrate simple uses of graphs and to build a general familiarity among students with graph models.

The problem-solving analysis required in this chapter is efficient, but ad hoc, logical examination of possibilities, a crucial part of the combinatorial reasoning in later counting chapters. Even small graphs are such complicated structures that brute force enumeration of possible subgraphs to, say, verify a particular graph property is normally infeasible. Mathematical insights are required to simplify and efficiently sort through the possibilities.

Section 1.1 Graph Models

The section quickly introduces the concept of a graph (vertices, edges, adjacency relation, and nothing else). Note that, as defined in this book, graphs cannot have loops or duplicate edges. Examples 6 and 7 are fairly involved; they present, in addition to novel graph models, further geometric models in turn for the graph models. These examples of interval graphs are a "pet" interest of the writer. Many other good examples of graph models may be found in the exercises.

Exercises Exercises 1-12 and 42-47 present other simple graph models. Exercises 13-41 mimic and extend the models introduced in the section's examples. Exercise 37 is a very neat application of interval graphs, although only better students will understand how to solve it.

Section 1.2 Isomorphism

This section examines the structure of graphs, via the problem-solving vehicle of asking whether two graphs are really different or are just "rearrangements" of each other. To answer such questions, it is necessary to begin introducing some graph terminology, such as, degree of a vertex and complete graph. Other terms, such as, circuit, are used informally and defined in later chapters.

The informal use of some graph terms and the arguments using these terms in Chapters I and II may be too lacking in rigor for some instructors. However the machinery needed to make all difinitions and proofs rigorous would greatly slow down the pace of the course and sidetrack the purposes of the course. Chapters III and IV have more rigorous presentations.

Exercises The graphs asked for in Exercise 13b, 13c and 14b do not exist.

Section 1.3 A Simple Counting Formula

The theorem and corollary in this section are the only general results known to apply to all graphs (by 'general', we mean without specially defined terms). The corollary is used to prove the non-existence of certain types of graphs, as demonstrated in Example 2.

Example 3 can be skipped in most courses. This seemingly contrived problem, known as Sperner's Lemma, can be used to prove Brouwer's Fixed Point Theorem and, more importantly, to find efficiently fixed points in economic equilibrium calculations.

Brouwer's Fixed Point Theorem can be stated: any continous function f(z) from a closed triangle onto that triangle has a fixed point x(i.e., f(x)=x). Sperner's Lemma can be used to prove this theorem by the following argument. Let the triangle T have its corners labeled with A 0, B 1, and C 2. Now triangulate T into tiny triangles, each with sides of length less than 1/n. Label a vertex z in this triangulation by the following rule: extend a line segment from f(z) through z to a point t on the edge of T, and now let the label of z be the label of the corner (A, B, or C) closest to t (break a tie arbitrarily). Sperner's Lemma (Example 3) says that one of the tiny triangles, call it T_n, has all 3 labels on its 3 corners. With a little analysis (i.e., real analysis), one can use the uniform continuity of f on the compact set T to prove that for $z_n \in T_n$, $|f(z_n) - z_n| < K/n$, where K is a constant depending on f. As $n \to \infty$, a subsequence of z_n's converges to a limit point z* for which f(z*) = z*. For details, see pages 8-10 in M. Todd's "The Computation of Fixed Points and Applications " (Springer-Verlag Lecture Notes in Economics and Mathematical Systems).

Exercises Exercise 6 is a curious, real-world application of the section's corollary.

Section 1.4 Planar Graphs

Planar graphs are of practical importance because they arise so often in operations research problems. Planar graphs are also of historical importance in graph theory because much of the development of the subject centered around the question, are all planar graphs 4-colorable. This section surveys some ways of determining whether a graph is planar. At first, a messy ad hoc geometric approach is used. Next Kuratowski's forbidden configuration is tried. Then some theory is developed: Euler's formula for planar graphs and a powerful corollary. From this corollary, many graphs can be shown to be non-planar with a simple algebraic test. This section is a case study in the power of a little good theory over ad hoc arguments.

Exercises The exercises develop several extensions of the theory and other basic concepts of planar graphs, e.g., properties of dual graphs in Exercise 12 and Platonic graphs in Exercise 25.

Supplement I Representing Graphs inside a Computer

This section presents the three most common ways to represent graphs in a computer. The material should be skipped unless time is available to assign some graph programming problems using these data

7

structures. Such programming is usually left to computer science courses.

Supplement II Supplementary Exercises

This section gives a survey of general graph theory problems. The problems examine a variety of graph terms, such as strong connectivity, cut-sets, and automorphisms. Exercise 4 uses the Pigeonhole Principle. Exercise 31 is a well-known Ramsey-theory question (Appendix 4); the solution (given in Part IV of this Manual) is famous for its elusive simplicity.

Chapter II   Covering Circuits and Coloring

This chapter presents three topics in graph theory, Euler circuits, Hamiltonian circuits, and graph coloring. These subjects are all over 100 years old, originally arose as games or semi-recreational mathematics, but have extensive uses today in operation research. Logical arguments are more carefully constructed and stated in this chapter, although only two theorems, the Euler Circuit Theorem and the Five Color Theorem, are actually proved. The 3 topics presented here are inherently appealing because of their recreational flavor. Further, what is fun is also seen to have practical applications. With this motivation, students are asked to perform the careful logical analysis of the possibilities required in ad hoc testing for Hamiltonian circuits and coloring. Section 2.4 on coloring theory can be skipped or covered briefly in most courses.

Section 2.1 Euler Circuits

Euler circuits are a pictorial topic with a simple, complete theory. This section presents an application to street sweeping and uses it to develop an alternate proof of the Euler circuit theorem.

Exercises There are several fairly easy variations of the Euler circuit theorem in the exercises: Exercises 6,7,8,13,19a and 20. Exercise 5 is equivalent to asking for a Hamiltonian circuit-- which does not exist for the given graph.

Section 2.2 Hamiltonian Circuits

The simple theory of Euler circuits is now replaced by the more normal graph-theoretic situation of complicated ad hoc arguments, here needed to determine whether a graph has a Hamiltonian circuit. The section ends with four theorems, stated without proof, about Hamiltonian circuits. Theorem 3 is not hard to prove and can be done in class.

<u>Exercises</u>   The graph in Exercise 4f (also appearing in Exer. 5
of section 2.1 and Exer. 1e in section 2.3) is called Petersen's
graph (it has many "bad" properties).   Exercise 6b has many
complicated subcases to consider in part ii. Exercises 8 and 9
present some "tricks" that can be used to prove quickly that certain
graphs cannot have a Hamiltonian circuit. Exercise 14 about a knight's
tour is a well-known result but finding the tour is tricky. The
following heuristic works: starting from any vertex, for the next move
always choose to go to the square from which the number of possible
subsequent moves is minimal. (Note': Ball and Coxeter's "Mathematical
Recreations and Essays " (U. of Toronto Press), p.175-184, has a
facinating discussion of the history of the knight's tour problem,
including contributions of Euler and DeMoivre).

Section 2.3 Graph Coloring

    This section presents logical ad hoc arguments for determining
the chromatic number of a graph and illustrates the practical uses
of coloring with examples in operatings research and computer science.
Although the garbage collection model takes a whole page to explain,
students have always found it very interesting.

    <u>Exercises</u>   The trick to showing the graph in Exercise 1g cannot
be 3-colored is first showing that a 3-coloring of the left half
(vertices a,b,c,d,e,f,g,h) forces a and h to have different colors
(as if they were joined by an edge), and similarly for i and p in the
right half. Exercise 8 is a good review of geography. Exercises 12,
13, 14 are simple color modeling problems.   Exercise 17 is a nice
practical use of coloring.

Section 2.4 Coloring Theorems

    Coloring theory, motivated by a century of research on the Four-Color
Problem, is perhaps the most extensively developed area of graph
theory. This section samples some basic results of this theory and
includes the "mandatory" theorem and proof of coloring theory, the
Five Color Theorem. The original false 1876 proof of the Four Color
Problem by Kempe used the same argument that works successfully in
the Five Color Theorem proof. Namely, in Figure 1 suppose e is
color 2 and G-x is 4-colored; try a 1-3 interchange at a and a
1-4 interchange at a; if both fail (there is a 1-3 path from a
to c and a 1-4 path from a to d), now try a 2-3 interchange at e
and a 2-4 interchange at b to make e 3 and b 4 and permit x to
get color 2: the error is that the 2-3 interchange at e may break
the 1-3 path from a to c so that the 2-4 interchange at b may also
change d to 2.

    <u>Exercises</u> Virtually all the exercises in this section are proofs.

9

Chapter III Trees and Searching

This chapter begins with some terminology and basic theorems about trees. The rest of the chapter applies trees to various searching problems. The trees are used to build search algorithms and to analyze these procedures. Section 3.2 introduces the two most common types of searches, depth-first search and breadth-first search. Section 3.3 illustrates two very different uses of trees in attacking complex operations research problems, the Branch and Bound method and heuristic approximation. The sample operations research problem chosen here is the famous Traveling Salesperson Problem. Section 3.4 introduces spanning trees and their role in graph algorithms. Section 3.5 uses trees to analyze the behavior of several sorting algorithms.

This chapter, possibly minus section 3.3, is essential material for computer scientists (although it may be covered in other computer science courses).

Section 3.1 Properties of Trees

This section begins with a lot of tree-related definitions. There are three theorems giving formulas involving various parameters of trees. Theorem 3 is used in section 3.5 to show that any algorithm for sorting n items needs at least $\log_2(n!)$ $(=O(n\log_2 n)$) comparisons in the worst case. Example 5 gives a mildly technical but very important use of trees in compiler design.

   Exercises   Exercises 2-15 are theory problems and exercises 17-27 involve applications. Exercise 13 is a bit tricky (see solution in Part IV).

Section 3.2 Enumeration with Trees

This section presents depth-first and breadth-first search methods with recreational examples. Example 4, involving a recursive program, should be skipped if most students have limited programming experience.

   Exercises   The later algorithm exercises (exercises 15-19) are very hard.

Section 3.3 The Traveling Salesperson Problem

This section presents the Branch and Bound technique, a tree-based procedure to search for optimal solutions in combinatorial operations research problems. The problem used to illustrate the technique is the famous Traveling Salesper. Problem. Such complicated operations research problems are often attacked with heuristic procedures that yield near optimal solutions. A tree-based heuristic for the Traveling Salesper. Problem is given in the second part of this section. The proof bounding the accuracy of this heuristic, while basically straightforward, is a bit difficult for students to follow and may be skipped.

Section 3.4 Spanning Trees and Graph Algorithms

This section discusses ways to build and traverse spanning trees. The absence of cross edges in depth-first spanning trees is noted and used to develop an algorithm for finding biconnected components in a graph. The biconnected components algorithm requires careful explanation but it is an example of a very important current topic, graph algorithms.

Section 3.5 Tree Analysis of Sorting Algorithms

Binary trees are a convenient model for depicting how most sorting algorithms work. They are also a necessary tool for analyzing the efficiency of most sorting algorithms. This section concludes with a sorting procedure using partially ordered trees, called heaps (rather than sort the whole list at once, a heap sort picks out one element at a time from a heap).

Although most computer science students will have studied this material before, this brief revisit with its tree approach should be valuable.

Exercises Exercises 4a,b and 7a, b are an instructive pair of exercises, showing the equivalence of two apparently different sorting procedures.

Chapter IV Network Algorithms

The theory of network flows is one of the great accomplishments in graph theory, both in terms of mathematical elegance and practical uses. This chapter first "warms up" for the augmenting flow algorithm with sections on shortest path and minimal spanning tree algorithms. The main part of this chapter is section 4.3 which introduces the basic theory of network flows. Section 4.4 applies flows to the combinatorial theory of matching. An important pedagogical aspect of network flows is the use of an algoorithm to prove a theorem (the Max Flow-Min Cut Theorem); the proof is literally constructed by the augmenting flow algorithm. This interplay of algorithms and theory is typical of much recent theory in computer science and operations research.

Section 4.1 Shortest Paths

Dijstra's shortest path algorithm is presented. It is a good example of operations research algorithms and serves to prepare students for the labeling procedure used in the augmenting flow algorithm in section 4.3.

Section 4.2 Minimal Spanning Trees

Minimal spanning tree algorithms are somewhat like Euler circuits;
the results in both are uncharacteristically nice. Actually, while
easy to state, minimal spanning tree algorithms do require some work
to verify. If time is tight, it is better to skip the proof of
Prim's algorithm and spend the extra time on section 4.3.

Section 4.3 Network Flows

The section begins with the terminology of network flows and
basic properties of flows and cuts. The presentation here is
mathematically rigorous. Corollary 2a should be carefully explained,
for it is the key to understanding the flow algorithm. Next an
intuitive but faulty approach to flow building is illustrated. This
is followed by the correct approach, the Augmenting Flow Algorithm of
Ford and Fulkerson. Finally, several basic variations of network flows
are modeled by standard single source-single sink flows.

Exercises Exercises 12 and 13 present another basic variation
on flows. Exercises 1-17 are computational problems; exercises 18-41
are theory problems.

Section 4.4 Algorithmic Matching

The theory of network flows is applied in this section to solve
matching problems and to prove the two fundamental theorems of
matching theory. Matching problems are an application of flows in
which negative labeling arises naturally. The theory in this section
should be skipped for students with limited mathematical maturity; the
first two pages of section 4.4 should then be treated as just one
more example in section 4.3.

Exercises Exercises 1-11 are computational problems; exercises
12-27 are theory problems.

Chapter V General Counting Methods for Arrangements and Selections

This is the most important chapter in the text. It introduces
combinatorial problem-solving. There is no theory, and only a few
basic formulas: just lots of examples to help prepare students to
do their own problem-solving. Some students will have difficulty
solving problems whose analysis does not mimic an example in the
text. It is virtually impossible to teach students the right way to
do problems in advance. Rather, much of the learning will occur
after assignments are done during homework discussion sessions.

As students worry about more complicated constraints found later in the chapter, they may become so uncertain of themselves that they no longer can solve easier problems. It is helpful to establish some personal rapport with students if uncertainty is setting in. Tell them this material is tricky for professors too(it is!) and encourage them to come to office hours with their questions. Students easily misinterpret problems, and even extensive discussion of an exercise in class may not clear up the confusion. Only after the student slowly explains the way he or she analysed the problem, can the instructor be in a position to help. If some students come to see the instructor with questions on the homework before the homework is discussed in class, then the instructor will be better prepared to help others with their questions during the class.

It is extremely important to stress the value of learning from one's mistakes. A good analogy is learning to throw darts. If you throw just one dart and it hits the bullseye, you were lucky but have learned little about controlling the flight of a dart. But if you start with a poor throw and successively improve your aim until you hit the bullseye, then you have really learned to throw darts.

If time permits, the instructor can preface this chapter with Appendix V on Mastermind-- with the intent of showing that combinatorial reasoning can be fun. The progression of increasingly more involved problems from Section 5.1 through 5.5 is the standard development found in most combinatorics texts. To lighten the possible tedium of combinatorial word problems, humorous settings are used in Section 5.1 and 5.2, e.g., Professor Mindthumper and names from Tolkien's "Lord of the Rings". The instructor may want to continue the levity by changing the setting of subsequent examples in the book, e.g., how many distributions are there of 20 identical administrators into...

Section 5.5 on binomial identities may be skipped. Section 5.6 on algorithms for combinatorial enumeration can be integrated with earlier sections so that programming exercises can be assigned throughtout this chapter. Linking combinatorial analysis with programming has an important dividend: it is often helpful to think of building a formula for a enumeration problem as a dynamic process, as if one were a computer that was printing out page after page of all possible outcomes. The fundamental question for the computer, what outcome to print next (e.g., what is the next outcome in this subclass, or is this subclass of outcomes completed and if so, what subclass should be listed next), generally requires the same type of analysis as needed to obtain a formula to count all outcomes.

Section 5.1 Two Basic Counting Principles

The Addition and Multiplication Principles, for breaking problems into disjoint or sequential subproblems, are the fundamental building blocks of all combinatorial enumeration. Later problems in this chapter will involve repeated, intermixed use of these two principles. Even though combinatorics is full of cute short solutions, the best way to first approach any counting problem is by using these two principles to produce a thorough case-by-case decomposition into small, manageable subproblems. Incomplete and incorrect decompositions are the cause of most counting errors.
Part d) of Example 5 illustrates a common error that should be emphasized   It is hard for students to anticipate such mistakes. Instead, they must learn from their mistakes. This is why a large amount of class time in the course, in this writer's opinion, should be devoted to discussing homework, i.e., discussing the faults in wrong solutions so that students will be able to avoid them in the future.

Exercises Exercises   12, 28, 30, 31, 35, 41 and 42 are primarily modeling or interpretation problems; the counting itself is easy once one knows what to count. Example 46 presents the game of Swap, with n white pegs and n blue pegs separated by a space; whites move left either one step (to an open space) or two steps (jumping over a blue peg to a following open space); blues move right in a similar fashion. For n=3, the game is played: B B B⁀_ W W W,

B B _B⁀W W W, B B W B _⁀W W , B B W B W⁀ _ W , B B W⁀ _ W B W ,

B⁀_ W B W B W , _B W B W B W (3 successive moves), W B W⁀B W⁀B⁀_,

W _W B W B B , W W W B⁀_ B B , W W W _ B B B. This game should be explained in class, before it is assigned.

Section 5.2 Simple Arrangements and Selections

Simple arrangements and selections are the small, manageable subproblems into which complex counting problems are commonly decomposed. This section introduces basic types of unconstrained and constrained arrangement and selection problems. The most common error in enumerating unordered sets is addressed in the Set Composition Principle and is demonstrated in Example 6c. This mistake of implicitly ordering the same set several different ways (picking a first part and then a second part of a set different ways) will be made over and over again in homework solutions.

Exercises The exercises in this section contain many different types of constrained arrangement and selection problems. The worked examples are of little help in attacking many of the later problems. Some students will get overwhelmed with the amount of ingenuity required and the variety of these exercises. Exercise 42 is the famous Birthday Paradox. Exercise 53 , counting poker hands with exactly one pair, is a pet problem of this writer; it has many possible right and wrong analyses. Exercises 50 and 51 introduce two important combinatorial problems in sampling theory.

Section 5.3 Arrangement and Selections with Repetition

Most students have seen the formula for arrangements with repetition, but the formula for selections with repetition is usually new. This section introduces problems involving simple constraints with repetition.

Exercises The first 15 exercises are similar to the examples. The remaining exercises involve more complex constraints and, in general, are quite difficult word problems. Many deal with arrangements in which certain consecutive pairs of elements may not appear. If such non-consecutivity exercises are assigned, it might be advisable to do one first in class.

Section 5.4 Distributions

This section presents equivalent distribution models for arrangements and selections with repetition. The section extends and compounds the examples in the previous section. There are more complex constraints, identical and distinct objects are intermixed, certain consecutive pairs are forbidden, and the model of integer solutions of an equation is introduced. The equation model is central to the development of generating functions in the next chapter. This section is the most difficult part of the course for most students. For several sections, students have been faced with increasingly complicated problems. At each stage, they had to move on without time to master fully the current problems. Assigning some review exercises from previous sections along with new problems is worth considering here.

Exercises Exercises 18-23 provide straightforward, yet valuable, practice in translating between different arrangement and selection models.

Section 5.5   Binomial Coefficients

This section begins with an explanation of how binomial coefficients get their name, i.e., how the $C(n,k)$'s arise in the binomial expansion. Then the symmetry relation and Pascal's recurrence relation are given. This is all material in this section that many courses will cover. The rest of the section concerns binomial identities. At the

end, the generalized binomial coefficients, for n any real number,
are introduced. Although binomial identities do not appear to be
problem-solving, the combinatorial reasoning used to verify them
has much in common with the reasoning used in previous sections.
Moreover, as shown in Example 4 and 5, identity (7) provides a
convenient way to evaluate combinatorial sums. This summation
technique is used in section 7.5 to solve a special class of recurrence relations.

Exercises Exercise 32 is used in Example 3 of section

Section 5.6 Generating Permutations and Combinations

This section gives algorithms for listing lexicographically
permutations, all combinations, and r-combinations of an n-set.
And it shows how to use these algorithms to enumerate outcomes in
earlier counting problems. This material can be integrated into
the preceding sections of this chapter. Writing out these
algorithms precisely in some programming language is important
for computer science students.
Exercises Exericses 16,17,18 about finding a given permutation's
or combination's position in a lexicographic list are good problems
for advanced students.

Chapter VI Generating Functions

Generating functions are probably the most important enumeration
model in combinatorics. In contrast to the unstructured problem-
solving of Chapter V , this chapter attacks problems with a very
specific, well-structured model, that of polynomial multiplication.
Section 6.1 explains how ordinary generating functions can model
problems of selection with restricted repetition. The polynomial
model is the sole message of this section. Section 6.2, on the other
hand, is concerned with techniques for evaluating coefficients in
given generating functions. These first two sections are the core
material of this chapter. Many courses will skip the rest of the
chapter. Section 6.3 looks at partitions and their generating
functions, as well as Ferrars graphs of partitions. Section 6.4
introduces exponential generating functions. The modeling performed
by these functions is more difficult to understand. In most undergraduate courses, it would be better to concentrate on a solid
mastery of ordinary generating functions. Section 6.5 shows how
summation problems can be solved with generating functions. This
section is a prerequisite for section 7.6 on generating function
methods for solving recurrence relations.

Section 6.1 Generating Function Models

In this section, students are shown how to build generating
function models for selection problems with restricted types of
repetition. The section starts with a review of the combinatorial
explanation of the Binomial Theorem given in section 5.5. However,

the Binomial Theorem can easily be explained at this point from scratch, if section 5.5 was skipped. Another polynomial expansion problem is restated as a combinatorics problem. Then the direction of the modeling is reversed, and combinatorial problems are restated as polynomial multiplication problems. The three key concepts in this section are: a) polynomial multiplication generates a set of formal products, as in expansion (3); b) these formal products are characterized as a sum of exponents; c) the number of such sums of exponents is just an integer-solution-of-equation problem, and conversely, any integer-solution-of-equation problem can be viewed as a sum of exponents problem. Any one of these three concepts may prove to be a major stumbling block for students. The best cure is for the instructor and a student to go slowly over the modeling process with a sample problem in both directions, from polynomial expansion to counting problem and from counting problem to polynomial expansion.

Exercises   The first 18 exercises are fairly straightforward. Exercises 20 and 21 are important examples of faulty generating functions. Even the very last exercises on multivariate generating functions are not hard.

Section 6.2   Calculating Coefficients

This section presents the necessary algebraic manipulation for evaluating particular coefficients of common types of generating functions. As these techniques are demonstrated, their combinatorial interpretations are also discussed. Letting rote algebraic manipulations do the work of the combinatorial analyses in Chapter V is an good example of the algebraic spirit of modern mathematics.

Exercises   Exercises 7 and 11 present variations of the expansion formulas given in this seciton. Exercises 38, 39, 41, 42, and 44 cover the basic properties of probability generating functions. Exercises 43 and 44 present examples of the "chain rule" for generating functions (exercise 44 is a basic result of queueing theory).

Section 6.3 Partitions

A brief discussion of partitions is given, primarily as an example of infinite generating functions. The section finishes with the Ferrers graph diagram for representing partitions and partition identities.

Exercises   Exercise 16 is an interesting use of Ferrars graphs, to prove that integer multiplication is commutative (this same problem is given as an induction proof exercise in Appendix 2).

Section 6.4 Exponential Generating Functions

This section begins with a careful discussion of a sample exponential generating function model. The fact that each formal product now contributes a value of $n!/e_1!\ldots e_r!$ to the coefficient makes exponential models much harder for students to understand. It is pointless to present this material unless the vast majority of the students already understand how ordinary generating functions work.

Exercises Exercises 21 and 22 continue the probability generating function exercises in section 6.2.

Section 6.5 A Summation Method

This section shows how to build a generating function with coefficient a_r a given combinatorial expression. Then the " summation operator" $1/(1-x)$ is used to sum such summands. The material in this section is used for solving recurrence relations in sections 7.5 and 7.6.

Chapter VII Recurrence Relations

A recurrence relation incorporates in a formal mathematical equation the combinatorial reasoning for decomposing a counting problem into similar subproblems. While recurrence relations are a fundamental tool in computer science for analyzing recursive algorithms, such relations have much wider applicability in all applied combinatorics than is generally realized. If enough parameters are used, virtually any counting problem can be modeled with a recurrence relation. This chapter's primary objective is teaching students how to build recurrence relation models. Its secondary (optional) objective is to present a brief survey of methods for closed-form solutions to recurrence relations. Solving the relations is of limited importance because: a) an applied combinatorics course normally emphasizes the modeling side of problem-solving; b) medium-sized problems are readily solved numerically by iteration, once a recurrence relation is available; and c) the solution techniques, with the exception of generating function methods, are of little general pedagogical value; one simply determines parameters in "cookbook" families of solutions.

Section 7.1 Recurrence Relation Models

This section is the heart of the chapter, and may be the only section some courses cover in this chapter. A variety of examples of recurrence relations are given. The section concludes with a discussion of the connection between recurrence relations and difference equations. This last material can be skipped; it raises an ancillary issue that can partially sidetrack students who should be concentrating on finding recurrence relation models. Like generating functions, recurrence models may confound some students at first. However, again a little private help should quickly straighten things out. A warning about compound interest relations: most students have yet to worry about savings accounts, mortgages, etc., and so they will not see the great practical importance in these relations that the instructor does.

Exercises Exercise 7 is the original Fibonacci problem about multiplying rabbits, the problem after which the Fibonacci relation is named; assigning it is a <u>must</u> (unless covered in class). Exercise 36 involves a product of the form $a_{n-1}a_0 + a_{n-2}a_1 + \ldots$ Exercises 37 and 45 are quite tricky. Exercises 23 through 30 involve primarily multi-variate relations and simultaneous relations. Students should be told that, when in doubt, it helps to add additional variables or supplementary equations.

Section 7.2 Divide-and-Conquer Relations

This section presents "cookbook" solutions to a class of recurr-ence relations that arise frequently in the analysis of recursive algorithms. Example 3 demonstrates, in a simple form, the idea behind Strassen's famous fast matrix multiplication (that multiplies two n x n matrices with less than n^3 multiplications).

Section 7.3 Recursive Programming

This section applies recursive reasoning to build computer programs that enumerate all outcomes in a combinatorial collec-tion. The section is a companion to Section 5.6. If most students do not have a good computing background, this section may need to be skipped.

Section 7.4 Solutions of Linear Recurrence Relations

Linear recurrence relations are the simplest common recurrence relations. Even if this section is not assigned, the instructor should spend a few minutes in class sketching the form of solution to linear recurrence relations. The key pedagogical points in this section are the similarity of the forms of solution for linear recurrence   relations and linear differential equations, and that the general solution to a linear recurrence relation is really a family of solutions, one for each possible set of initial conditions. Example 4 is fairly complicated and the instructor may want to skip it (to avoid scaring students).

Section 7.5 Solutions of Inhomogeneous Recurrence Relations

This section presents solution techniques for inhomogeneous first-order linear recurrence relations:
$$a_n = ka_{n-1} + f(n). \quad \text{If } k \neq 1, \ a_n$$

will be essentially the same type of function as f(n). If k = 1, the relation is just a summation problem. In the latter case, it is nec-essary to have previously discussed summations, either using binomial identities (section 5.5) or generating functions (section 6.5).

Section 7.6 Solutions with Generating Functions

This section shows how recurrence relations can be converted into equations for associated generating functions. This conversion is the discrete   counterpart to transform techniques for solving differential equations. While important, this material is very difficult for most undergraduates who are not upper-division math majors. It could require upwards of a week of the course to present this section properly to other students. Examples 2 and 5 use partial fraction decompositions (a technique that, at best, students saw briefly in calculus and quickly forgot); to determine coefficients in partial fraction expansions, consider using the Heavyside method (discussed in Thomas's <u>Calculus</u>).

A concrete application of order of multiplication (Example 4) is matrix multiplication, where the number of simple entry-times-entry multiplications in dependent on the placement of parentheses.

Chapter VIII   Inclusion - Exclusion

In this chapter, the counting principle of complementation, that the number of objects without property A is the total number minus the number with property A, is generalized to situations involving many properties. Many problems presented in Chapters V. and VI can be solved more easily now. The "theory" of this chapter consists of working out in general set-theoretic terms a formula, expressed in terms of the sizes of various set intersections, for the number of objects that have none of a group of properties. To solve a given problem with this formula, a student must: a) define a group of properties; and b) count the number of objects with various subsets of the properties-- these latter counting problems are usually simple Chapter V-type counting problems. The last section develops a nice mini-theory for solving arrangement problems with restricted positions. This material is not difficult, but time constraints may force this section to be skipped.

Section 8.1 Counting with Venn Diagrams

This section slowly generalizes the principle of complementation from one property to two properties and finally to three properties. Students often have trouble even in the 2-property case. First, they have difficulty defining the properties that are not to hold for the elements to be counted. This is a matter of logical negation in set theory and propositional logic (i.e., $\overline{A \cap B} = \overline{A} \cup \overline{B}$).   A second problem is trusting in a set-theoretic formula.   In Chapter V , similar problems had to be carefully thought out and broken into appropriate subcases. Now students are conditioned to such an approach rather than "plugging into" a formula.

Exercises Exercises 25-29, like Example 6, cannot be solved by inclusion-exclusion, but rather require ad hoc Boolean algegra analysis. Exercises 17, 21, and 22 ask for $N(A_1 \cup A_2 \cup A_3)$ instead of $N(\bar{A}_1 \bar{A}_2 \bar{A}_3)$. The counting subproblems in Exercises 12,21, 22, and 23 are moderately complicated.

Section 8.2 Inclusion-Exclusion Formula

Now the students should be ready for the general Inclusion-Exclusion formula. Some students will already have forgotten the necessary Chapter V-type counting methods needed for solving some of the subproblems arising in the use of the Inclusion-Exclusion formula. A few students may be bothered by the generality of the summation notation in the formula. The proof of the Inclusion-Exclusion formula need not be given; it has already been motivated by the discussion in section 8.1. Theorem 2, at the end of the section, can be skipped for most audiences.

Exercises Exercise 13 is a trick; it is really a selection-with-repetition problem with at least one of each type (Inclusion-Exclusion is not needed). Exercise 21a looks like a derangement problem but is not (there are n-1 properties, not n properties). Exercise 24 is a tricky "recursive" Inclusion-Exclusion problem (see its solution in Part IV of this Guide).

Section 8.3 Restricted Positions and Rook Polynomials

This section on rook polynomials presents a clever model for restricted position problems using chessboards and generating function models. Each step of the theory is easy to follow but the final result is far from obvious. The case of non-disjoint subboards (starting after Example 1) may be skipped.

Exercises Exercises 2c,d,e, 5b, and 6 do not decompose into disjoint subboards (but 6 is not hard to do by inspection).

Chapter IX Polya's Enumeration Formula

This chapter discusses a counting problem in applied group theory, a problem in which generating functions play an important role. This chapter has the only extensive theoretical development in the enumeration half of this book. The basic definitions and theoretical development are presented by examples without formal proofs. The emphasis is on understanding the theory by using it to solve problems. One of the pedagogical goals of this chapter is to present groups in a practical and natural setting (as opposed to the more formal setting in a modern algebra course). Polya's formula is usually not discussed in sophomore-junior-level applied combinatorics courses, because of its abstract foundations in group theory. However, the presentation "by example" used here is much easier to follow than the standard treatment. While the students at this writer's school have trouble following the explanation of Theorem in section 9.2, they have no particular difficulty with the subsequent problems based on the Theorem.

Section 9.1 Equivalence and Symmetry Groups

This section introduces the sample problem of two colorings of the corners of a square. It is most helpful to have a physical model (say, of tinker toys) of a square to show how motions transform one coloring pattern of the square's corners into another pattern. The concepts of an equivalence relation and a group are presented and linked together. It is important here to emphasize the difference between a motion's permutation of the corners of the square and a motion's induced permutation of the colorings of the corners-- the latter permutations generate the equivalence classes we want to count. Examples 2,3, and 4 about enumerating symmetries are best left as out-of-class reading (or may be skipped).

Exercises   Exercise 4 is a valuable warm-up for cycle decompositions in sections 9.3 and 9.4. Exercise 12 on the non-commutativity of symmetries should be assigned or covered in class. Exercises 16-25 involve basic group theory.

Section 9.2 Burnside's Lemma

The section on Burnside's Lemma is the most difficult section in the book. A motivating argument for this algebraic lemma is attempted. If most students have previously had a modern algebra, a proof of the lemma, as in C.L. Liu's "Introduction to Combinatorial Mathematics", should be given. Otherwise, the informal argument should be discussed for 20 to 30 minutes (using a physical-model of the square to illustrate the C_{10} example) and then the instructor should turn to applications of the lemma's formula. The idea of counting colorings with "multiplicities" can be mentioned while doing example 1. Students should not worry about remembering the explanation of the lemma.

Exercises   Exercise 11 is a cute (and tricky) combinatorics problem. Exercises 13-16 continue the theoretical development of the theory exercises in section 9.1.

Section 9.3   The Cycle Index

The section is a self-explanatory development of the role of the cycle structure representation of a motion in counting the number of colorings left fixed by a motion. It is desirable to get maximum student participation in filling in the various entries in the table. The examples in this section are easy for students to read out of class; if possible, cover one in class.

Exercises   Exercise 8 presents a constraint most easily handled with the Inclusion-Exclusion formula.

Section 9.4

This section extends the development in section 9.3 to obtain the final Polya's formula. If Example 4 is discussed, a physical model of the cube is desirable.

Chapter X

This chapter uses the concepts of formal languages, finite-state machines and propositional models from theoretical computer science to look at combinatorical enumeration and optimization from a different point of view. At the same time, students are introduced to these important ideas without requiring extensive background in computer science.

The first two sections on languages and machines form a unified package, although section 10.1 could be presented alone. The last section on computational complexity can also be presented alone (although the definition of NP requires knowing what a Turing machine is).

A one-semester or one-quarter course in discrete structures for computer scientists should cover this chapter. A combinatorics course (with no graph theory), should cover the first two sections. Other courses probably will not have time for this chapter.

Section 10.1 Formal Languages and Grammars

Formal grammars provide a bit of algebraic formalism for the combinatorial enumeration done in the second part of this text. The concept of a regular grammar should be stressed because of the central role it plays in the next section.

Exercises The last three exercises on non-existence of grammars for certain languages are good problems are better students.

Section 10.2 Finite-state Machines

Finite-state machine design involves as much computer science as combinatorial analysis. The theorem relating regular languages and languages recognized by finite-state machines is pure computer science. Our construction in the proof of this theorem (actually just one half of the proof) is less complicated than the constructions in computer science texts.

Exercises Exercise 4 is a must.

Section 10.3 Logical Propositions and Computational Complexity

NP-completeness is one of the most powerful ideas to come along in some time. Showing that a problem cannot be solved in polynomial time is an awesomefully difficult task. The theory of NP-completeness allows one to measure the difficulty of one problem against well-known hard problems. The best thing about NP-completeness, from this book's viewpoint, is that the theory is all combinatorial modeling and reformulation.

Exercises After the first six problems, these exercises are quite hard.

Chapter XI Games with Graphs

This chapter develops the theory of progressively finite games and applies this theory to find winning strategies in certain 2-person games. The ultimate goal is a winning strategy for the game of Nim. This chapter starts with a self-contained graph-theoretic model for solving the Instant Insanity puzzle. For many years, this author has closed his course (i.e., last half-hour of the last class) with this clever graph-theoretic solution to Instant Insanity. The successful solution always evokes appause and ends the course on a high note.

The two sections on progressively finite grames have the most abstract material in this text. The definitions and proofs are all based on recursive constructions and thus have substantial pedagogical value, especially for computer scientists.

Section 11.1 Graph Model for Instant Insanity

This clever model speaks for itself. There are enough pieces to the model that a good 25-to-30 minutes should be allocated to present the model and its analysis. Rather than using the sample cubes in the text, the instructor should buy his or her own set of cubes and solve them in class. To keep the cubes distinct, number a corner of each cube 1 thru 4. Then practice arranging the cubes into the correct pile a few times before class (it is easy to make a mistake in arranging the cubes and end up triumphantly displaying an incorrect solution).

Section 11.2 Progressively Finite Games

This section introduces the basic concepts of progressively finite games: kernels, levels, and Grundy functions. The recursive constructions need to be constantly illustrated with specific examples.

Exercises Exercises 12 and 13 contain two important results (needless to say, the proof in Exercise 13 involves a recursive construction).

Section 11.3 Nim-type Games

This section introduces the complicating generalization of direct sums of games, and with it the operation of digital sum. Notation gets quite involved and the proof of the Theorem is beyond the grasp of most undergraduates. However by learning how to play winning Nim, they should get a little feeling for the idea behind the proof.

Exercises Exercise 4 is a concrete example of part b (the omitted part) of the proof of the Theorem.

APPENDICES

Appendices 1 and 3 have background material on set theory and probability. Appendix 2 presents induction, which should be review material but unfortunately is not for many students. Some instructors may want to spend one or two classes in an expanded treatment of induction. Appendix 4 presents the Pigeonhole Principle, an important topic in combinatorial theory. Appendix 5 presents the game of Mastermind, which can be used to introduce enumeration.

Appendix 1. Set Theory and Logic

This section presents basic set-theoretic terms, notation, and operations. The set complementation laws (deMorgan's Laws) should be stressed, since they will be used many times in later chapters. Example 1 illustrates the 'real-world' issue of how much information is needed to solve a counting problem. This problem, like the problem of counting imprecise sets, is raised as a reminder to students that questions must be well posed before one can employ the combinatorial reasoning taught in this text.

The brief restatement of sets and Boolean algebra in terms of propositions and propositional calculus is given for completeness. Most students prefer to think solely in terms of sets and Venn diagrams. The text uses only set-theoretic terminology in subsequent chapters.

If this text is being used in a Discrete Structures course, the instructor may want a more extensive discussion of set theory, propositional logic, and related topics, such as set functions. The instructor should consult texts such as F.P. Preparata and R.T. Yeh, "Introduction to Discrete Structures", Addison-Wesley (Reading, Mass., 1973) or L.L. Dornhoff and F.E. Hohn, "Applied Modern Algebra," Macmillan (New York, 1978).

Exercises Note that in exercise 2, some of the requested subsets cannot be generated (parts d and e). Exercise 5a has contradictory data.

Appendix 2 Mathematical Induction

Anyone who takes a course in discrete mathematics should be exposed to induction arguments. While students may be asked to do few induction proofs themselves in this course, they at least should be developing a familiarity with induction arguments for later courses in computer science and other areas of discrete mathematics. If an instructor wants to put more emphasis on induction, extra examples (chosen from the exercises) can be done in class, and one or two induction exercises from this section can be assigned every week for the first month. Induction arguments are used in this text to prove Euler's formula for planar graphs (section 1.4), the 5-color theorem (section 2.4), a bound on the height of m-ary trees (section 3.1), the validity of Prim's minimal spanning tree algorithm (section 4.2), and several theorems in Chapter XI Games with Graphs.

Exercises   Exercise 16 proves the commutativity of integer multiplication, a well-known fact that few can prove (this fact also has a simple combinatorial proof; see exercise 16 in section 6.3). Exercises 25 and 26 are classic examples of faulty induction arguments.

Appendix 3   A Little Probability

The objective of this section is to present the definition of probability used in this text, namely, the fraction of 'favorable' outcomes. However, in giving this definition, we immediately raise the problem of distinguishing outcomes, or elementary events, from compound events. This problem becomes fairly subtle when identical objects are involved. A point that can cause some confusion is that in probability problems, objects are always distinguishable (even if they have the same shape and color), whereas in combinatorics one allows identical objects. Note that identical objects arise implicitly in some probability problems, such as counting all sequences with k (identical) heads and n-k (identical) tails when a coin is flipped n (distiguishable) times. Probabilities are used in Chapter V to give interesting settings to counting problems. Some advanced exercises in section 6.2 and 6.4 examine basic properties of probability generating functions.

Exercise   Exercises 14 and 15 require students to identify elementary events in various experiments.

Appendix 4   Pigeonhole Principle

This brief section on the Pigeonhole Principle is often skipped, because there is no material in the rest of the text that uses (or reenforces) the principle. The idea can be explained in half a minute in any subsequent course that may need it. However it is a traditional combinatorics topic whose advanced theory has blossomed in recent years.   Some instructors may find this section appealing because " Pigeonhole" exercises require a type of creative abstract reasoning generally missing in the rest of the book. For a fuller, yet still elementary, discussion of the Pigeonhole Principle, see R. Brualdi, "Introduction to Combinatorics," North Holland (New York, 1975).

Exercises   Exercise 18 is a famous Ramsey-theory question with a simple, yet hard to discover, proof.

Appendix 5 Game of Mastermind

Mastermind is an enjoyable combinatorial game that requires the same disciplined combinatorial reasoning found in many subsequent combinatorial problems in this book and in real-world applied combinatorics. The case-by-case analysis of possible solutions will be repeated in other settings throughout Chapter II. The following device has proven quite popular with students: put a Mastermind problem (say, taken from the problems in Ault's book), laid out on a Mastermind board, in a glass display case each week. Such a display is good "publicity" for the course.

The real challenge for students in doing Mastermind exericses is to find simple analyses that make it possible to explain one's solution concisely. D. Knuth has published a complete analysis of an optimal way to play Mastermind (J. of Recreational Math. vol. 9 (1976), p.1-6).

Exercises Exercise 21 gives an easy approach to getting the Secret Code by the sixth guess (this exercise takes some of the fun out of the game and so use caution in assigning it).

Part III Sample Course Syllabi, Assignments and Tests
   (solutions to exercises may be found in Part IV of Manual)

1. One-semester Undergraduate Course emphasizing Graph Theory
   or One-quarter Graph Theory Course (using first ten weeks)

Week I- Reading 1.1, 1.2
   Exercises 1.1 #1,2,7,15,29ad,33,46// 1.2 #1,2,5,6,10a

Week II- Reading 1.3, 1.4, Supplement 1
   Exercises 1.3 #1,2,3,4,6// 1.4 #1a,2,3abc,8,14,17// Suppl.2 #5,19,24

Week III- Reading 2.1, 2.2
   Exercises 2.1 #1ab,2ab,3,4,5,15,16// 2.2 #1ab,2ab,4ab,6a,10,14

Week IV- Reading 2.3, 2.4
   Exercises 2.3 #1abg,2a,8,9ac,11,17// 2.4 #1,4,8,14

Week V- Review & Test ; also Suppl. 2 (of Chap. I) #31

Week VI- Reading 3.1, 3.2, 3.3
   Exercises 3.1 #1,4,5,6,17// 3.2 #2,10// 3.3 #1,5,7a

Week VII- Reading 3.4, 3.5
   Exercises 3.4 #1a,4,16,21,24,28,32a// 3.5 #4,7ab,13

Week VIII- Reading 4.1, 4.2, 4.3 (up to Max Flow-Min Cut Alg.)
   Exercises 4.1 #2ab,4// 4.2 #1a,2a// 4.3 #2abc,4d

Week IX- Reading 4.3, 4.4
   Exercises 4.3 #8,9,12,23a,24,29// 4.4 #1ab,6,8,11,13

Week X- Review & Test

Week XI- Reading 5.1, 5.2, Appendix 5
   Exercises 5.1 # 1,6,15,30,46// 5.2 #3,7,11,13,17,28// App. 5 #1,3

Week XII- Reading 5.3, 5.4
   Exercises 5.3 # 1,2,8,10,11,13,27// 5.4 #1,2,6,18ab,19ab,21ab,45

Week XIII- Reading 6.1, 6.2
   Exercises 6.1 #2,3ac,4ab,10,13,15,23// 6.2 #1,4,5,10,16,22,24

Week XIV- Reading 7.1, 7.2, 7.3
   Exercises 7.1 #1,3ab,5,6a,7,9,12// 7.2 #2,4// 7.3 Programs for
                                                      7.1 #3,6a
Week XV- Reading 8.1, 8.2
   Exercises 8.1 #2,9,10,11,14,29// 8.2 #1,5,7,17,19

Final Exam- covering last five weeks only

1. Are the following two graphs isomorphic? If so, label the vertices of the graph on the right to exhibit the isomorphism; if not, carefully explain why not.

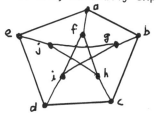 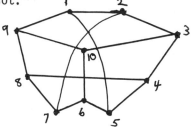

2. Show that a planar, polygonal graph with fewer than 30 edges must have at least one vertex with local degree less than or equal to 4.

3. (Note: for full credit on this problem, you must explain or draw a simple graph to illustrate your answer.)
   Consider a connected graph G which has no vertices with local degree $\rho = 1$. If the vertices of G can be properly two-colored:

   a. Must G have a Hamilton Path?

   b. Must the dual G* have an Euler Circuit?

   c. Characterize all graphs which are both face 2-colorable and vertex 2-colorable.

4. For the graph shown below:
   a. Find a proper minimum <u>vertex</u> coloring of the graph.

   b. Remove the edge marked with the X and then find a proper minimum <u>face</u> coloring of the graph.

   c. Draw the dual of the graph which you face-colored in part b.

   d. Find a proper minimum face coloring of the graph in part c.

5. The figure shown to the right is a triangular prism. It is a solid object whose bases are equilateral triangles and whose sides are rectangles, which are perpendicular to the planes of the bases. The prism is unoriented in three dimensions, and the vertices (corners) are to be 2-colored in all distinct ways. Make use of the Pattern Inventory to determine the number of colorings with two corners of one color and

2nd Test

1. **Suppose** we use the following search procedure to test which letter (A,B,C,D,or E) is the unknown letter X. First see if X=A, then if X=B, etc. draw the associated binary testing tree.

2. **Build** depth- 1st and breadth-1st spanning trees rooted at a for the graph on the right.

3. **Draw** the splitting and merging trees for a merge sort of the list 9,2,3,7,1,10,6,8,11,5,4.

4. **Perform** both a branch-and-bound and a Quick Construction Traveling Salesman analysis for the cost matrix on the right

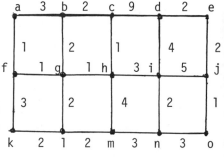

	1	2	3	4
1	∞	1	3	2
2	1	∞	4	2
3	3	4	∞	1
4	2	2	1	∞

5. For the network on the right find

   a) a shortest path from a to o that uses edge (c,d)

   b) a minimal spanning tree that includes edge (c,d)

Note that your answers will be minimal only with respect to this constraint, not globally shortest or minimal.

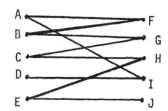

6. **Sup ose** the matching A-I, B-F, C-G, E-H has been made in this bipartite graph. Make the associated network and flow corresponding to this matching. Now apply the Augmenting Flow Algorithm to find a complete matching of all vertices.

1. Find the coefficient of x^{55} in $g(x) = [x^5 + x^8 + x^{11} + x^{14}]^5$.

2. How many ways can the letters PROBLEMS be arranged so that the P is not first nor the S last?

3. How many simultaneous integer solutions are there to the two equations, with the $x_i \geq 0$; i=1,2,3,4,5,6:

$$x_1 + x_2 + x_3 + x_4 + x_5 + x_6 = 20 \text{ and } x_1 + x_3 + x_5 = 7$$

4. How many arrangements are there of the letters INSTRUCTOR with the vowels separated from one another by exactly 2 consonants.

5. Find a recurrence relation for a_n, the number of ways for an image to be reflected n times by internal surfaces of two panes of glass. The diagram below shows that $a_0=1$, $a_1=2$, $a_2=3$. Explain your answer.

6. Twelve suitcases are checked in an airport. The suitcases all look alike, but belong three each to four men. When the men claim their suitcases the claim checks become confused. Find the number of ways in which each man can get three suitcases but with no man getting all three of his own.

2. One-Semester Undergraduate Course emphasizing counting or One-Quarter Enumerative Combinatorics Course (using first 10 weeks)

Week I- Reading Appendices 1, 2, 5
   Exercises App.1 #1ac,2ab// App.2 #1,8,14// App.5 #1,3

Week II- Reading 5.1, 5.2
   Exercises 5.1 #4,9,11,12,24,30// 5.2 #4,7,11,12,53a// App.2 #13

Week III- Reading 5.3, 5.4
   Exercises 5.2 #29a-i// 5.3 #1,6,9,10,17// 5.4 #3ac,7,9,11,23//
                                                        App. 2 #4

Week IV- Reading 5.5 (skip last 2 pages), 6.1
   Exercises 5.2 #27,46// 5.4 #16,19a,20a// 5.5 #1a,2a,11a// 6.1 #2,4,9,
                                                              13,19

Week V- Reading 6.2, 6.3
   Exercises 6.1 #16// 6.2 #2,4,8a,14ab,18,22,27// 6.3 #3,5,14

Week VI- Review and Test

Week VII- Reading 7.1, 7.2
   Exercises 7.1 #1,3,5,6a,7,10,12,41,22,26// 7,2 #2,9

Week VIII- Reading 8.1, 8.2 (skip last 2 pages)
   Exercises 8.1 #11,13,21,23,26// 8.2 #7,10,15,20,29b

Week IX- Reading 8.3, 9.1
   Exercises 8.2 #5,14,27// 8.3 #5ab,6// 9.1 #1,2,4,10,12

Week X- Reading 9.2, 9.3, 9.4
   Exercises 9.2 #10// 9.3 #1,3,6// 9.4 #2,4,5,8

Week XI- Review and Test

Week XII- Reading 1.1, 1.2
   Exercises 1.1 #2,6,9,13b,16b,20,22,29ab,37// 1.2 #5abc,14

Week XIII- 1.3, 1.4, 2.1
   Exercises 1.3 #1b,2ab,4,6// 1.4 #2,3,19// 2.1 #1,4,5,14,15

Week XIV- Reading 2.2, 2.3, 3.1
   Exercises 2.2 #4g,6a,7// 2.3 1abg,8) ,13// 3.1 #2;3,5,8a,12,20

Week XV- Reading 3.2, 3.3, 3.4
   Exercises 3.2 #4a,10,11// 3.3 #5,7b// 3.4 #4,6,11,24

Final Exam- covering graph theory only

1. Find the coefficient of x^{30} in $(x^3 + x^4 + x^5 + x^6 + x^7)^6$.

2. Model the following problem: distribute 13(identical) dimes among 4 children with 3 children getting at least 1 dime and the fourth child (the oldest) getting at least 2 dimes.
   a) state an equivalent selection with repetition problem.
   b) Model as a specified coefficient of a generating function.
   c) Solve this problem (any way you want).

3. Answer one of the following two questions:
   I. Give a generating function for the ways to partition an integer r into an (unordered) sum of integers, each integer appearing an odd number of times or not at all.
   II. Explain the following identity by either a combinatorial argument or by block-walking
   $$\binom{n}{0} + \binom{n}{1} + \binom{n}{2} + \ldots + \binom{n}{n} = 2^n$$

4. What is the probability that a random number between 0 and 9,999 has exactly one 8 and one 9?

5. Determine the number of non-negative (≥ 0) integer solutions to the equation
   $$2x_1 + 2x_2 + 2x_3 + x_4 + x_5 + x_6 = 7.$$

   Hint: break into cases based on the sum of the first 3 variables.

1. Give an expression for the number of ways to match the 5 men with the 5 women with the restrictions shown at the right.

2. Give the cycle structure representation for the following symmetry of the corners of a 10-gon: a flip around an opposite pair of corners.

3. Give a recurrence relation for a_n, the number of ways a sequence of 1's and 2's can sum to n (e.g., $a_3 = 3$: 111, 21, 12). Find a_8.

4. Give an expression for the pattern inventory for the number of ways to color the 8 edges black or white in the figure on the right that is unoriented in 3 dimensions (rotations and flips allowed).

Draw 2 black-white edge colorings equivalent in 3 dimensions but not in 2-dimensions.

5. How many rearrangements of the sequence   A,A,B,B,C,C,D,D,E,E are there in which at least one A is in a new position in the sequence, at least one B is in a new position, etc. (similarly for C, D, and E)?

Final Test

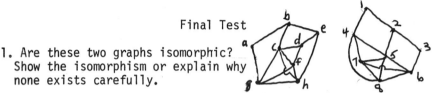

1. Are these two graphs isomorphic? Show the isomorphism or explain why none exists carefully.

2. Draw a (connected) planar graph, if possible (or explain why none exists), with: a) 7 vertices and 5 regions; b) 7 regions and all vertices of degree 4.

3. Suppose that the losers in the first 2 rounds of a tennis tournament with 32 entrants qualify for a losers tournament. The people who lose in the first two rounds of this tournament qualify for another tournament, etc. until finally there is a grand loser (the loser in a 2-person tournament). How many tournaments are required to determine this grand loser (assume all tournaments are balanced trees; if a person has a bye in the 1st round, then their initial (2nd rd.) match is the person's only chance to qualify for the next losers tournament).

4. Draw a non-planar 7vertex graph with no Hamiltonian or Euler circuits. Explain why your graph has none of these three properties.

5. In a round-robin tournament, where each pair of n contestants play each other on successive days, a major problem is scheduling the matches over a minimal number of days.
   a) Restate this scheduling problem as an <u>edge</u> coloring problem for an associated graph. Define what you mean by a proper edge coloring so that an edge coloring will correspond to a scheduling of matches in the tournament (who plays on which days; a person plays at most once on a single day).

   b) Find a minimal (minimal number of days) schedule for a 6-contestant tournament, and a minimal schedule for a 5-contestant tournament. Explain in each case why fewer days will not suffice.

34

3. One-Semester Graduate Course

Week I- Reading 1.1, Appendices 2, 4
  Exercises 1.1 #9,16bc,30,31,34,37,40,43// App.2 #4,11// App.4 #4,10,14

Week II- Reading 1.2, 1.3, 1.4
  Exercises 1.2 #5bd,7,14// 2.3 #2,4,6,7// 1.4 #3ace,6,16,19,21

Week III- Reading 2.1, 2.2, 2.3, 2.4
  Exercises 2.1 #2,14,15,16// 2.2 #4a,6b,7a,14// 2.3 #1bg,8,9,11//
                                                    2.4 #4,6
Week IV- Reading 3.1, 3.2, 3.3
  Exercises 3.1 #4,5,6,13,20,27// 3.2 #10,11// 3.3 #5,7c

Week V- Reading 3.4, 3.5, 4.1, 4.2
  Exercises 3.4 #1b,11,14,24,29,32abc// 3.5 #4,7ab,13abc// 4.1 #4,11//
                                                    4.2 #1a,5,9,17
Week VI- Reading 4.3, 4.4
  Exercises 4.3 #1,4d,8,9,12,25// 4.4 #1,8,11,13,15

Week VII- Review & Test

Week VIII- Reading 5.1, 5.2, Appendix 5
  Exercises 5.1 #4,11,25,30,35,40// 5.2 #11,28a-i,36,50,53ab,54//
                                                    App.5 #2
Week IX- Reading 5.3, 5.4, 5.5, 5.6
  Exercises 5.3 #9,19,29// 5.4 #8,11,20,35,45,50// 5.5 #2bc,11a,29c//
                                                    5.6 #4,22
Week X- Reading 6.1, 6.2, 6.3, 6.4
  Exercises 6.1 #8,13,16,20,27// 6.2 #13a,18,22,38c,39ab,43//
                                                    6.4 #5,8,11
Week XI- Reading 7.1, 7.2, 7.6
  Exercises 7.1 #3,6,13,14,22,26,36,40,45// 7.2 #5// 7.6 #1a,7

Week XII- Reading 8.1, 8.2, 8.3
  Exercises 8.1 #11,17,22,28,29// 8.2 #7,10a,20,24,27// 8.3 #2d,5,6

Week XIII- Reading 9.1, 9.2, 9.3, 9.4
  Exercises 9.1 #1,10c,12// 9.2 #3,9,11,13a// 9.3 #3,4d,13//
                                                    9.4 #2bd,9b
Week XIV- Reading 10.1, 10.2, 10.3
  Exercises 10.1 #5,11,13// 10.2 #5,11,17// 10.3 #4a,5,11

Week XV- Review

Final Exam- covering second half of course

First Test

1. a) Find a depth-first spanning tree for the directed graph on the right.
   b) Index the vertices according to a preorder traversal of this spanning tree.
2. The matrix on the right has a 1 in entry (i,j) if boy B_i is compatible with girl G_j. Suppose we want to pair each boy and girl with 2 compatible people. A partial 2-pairing is indicated with circles.

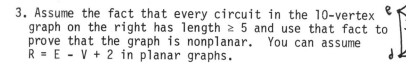

   a) Model this problem with a network (indicate all edge capacities) and build a flow corresponding to the given partial 2-matching.
   b) Use the Augmenting Flow Algorithm on this network. Show labels & final 2-matching.

3. Assume the fact that every circuit in the 10-vertex graph on the right has length ≥ 5 and use that fact to prove that the graph is nonplanar. You can assume $R = E - V + 2$ in planar graphs.

4. Prove carefully that the graph in problem 3 has no Hamiltonian circuit.

5. Prove by induction that the minimum number of colors required to edge color a tree T (edges with a common vertex must be different colors) is equal to the maximum degree of a vertex in T.

6. Find a graph G that cannot be modeled by any family of overlapping arcs on a circle but such that every subgraph of G (obtained by removing one or more vertices) can be so modeled. (In a circular-arc model, there is a one-to-one matching of vertices of G with arcs such that a pair of vertices are adjacent if and only if the corresponding arcs overlap.)

Final Test

1. Give an expression of the rook polynomial for the board on the right.

2. Find the coefficient of x^{45} in $(x^3 + x^4 + x^6 + x^7 + x^8)^7$.

3. Build a generating function model for the following problem: How many n- digit telephone numbers are there in which every digit (0 or 1 or ... or 9) that appears occurs at least twice.

4. Give an expression for the pattern inventory for the 3-colorings (black, white, red) of the 7 <u>edges</u> of the unoriented (in 3 dimensions) figure on the right.

5. If a pair of identical dice are rolled n successive times ($n \geq 6$), how many sequences of outcomes of the n rolls are there that contain all doubles (a pair of 1's, a pair of 2's,..., a pair of 6's; note that each pair must occur one <u>or more</u> times).

6. Find a recurrence relation for a_n, the number of ways to arrange the integers 1, 2,..., n in a sequence such that each integer in the sequence differs by 1 from some integer following it in the sequence (ignore the constraint for the last integer). Explain your answer.
   Hint: first determine a_1, a_2, a_3, a_4 by inspection.

7. How many arrangements are there of 3 a's, 4 b's, 5 c's, and 6 d's in which each d is beside another d?

37

Part IV Solutions to Selected Exercises

Chapter I

Section 1.1

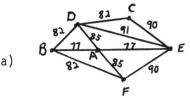

1. a)

b) 2, e.g. (C,D) and (C,E), c) yes $C \overset{82}{\rule{1em}{0.4pt}} D \overset{91}{\rule{1em}{0.4pt}} E \overset{90}{\rule{1em}{0.4pt}} F \overset{85}{\rule{1em}{0.4pt}} A \overset{77}{\rule{1em}{0.4pt}} B$

2. a) [diagram] b) ADCB, BADC, CBAD, DCBA, DBAC

3. a) [diagram], b) consider 2 teams say, A and B, that play each
   other: they cannot play their other games against the same
   third team or else the remaining two teams would have to play
   each other twice. So A and B have different other opponents,
   call them C and D. The remaining team, call it E, can now only
   play C and D.

4. a) [diagram] b)

5. a) Vertex = Board, edge joins Boards with overlap,
   b) 6:a,d,f,g,i,k by inspection - break with two subproblems:
   subgraph of a,b,e,f,i,j (max 3) and subgraph of c,d,g,h,k (max
   3).

6. a) J [diagram] W, b) 2 days, c) Yes (J,W) or (J,M).
   M R S T

7. a) [diagram] (b,c) and (j,k) must have opposite directions and each vertex must have (at least) one inward edge and
   one outward edge, b) Min = 8: from a to c or from c to a,
   depending on how (b,c) is directed.

8. 2 edges: (b,d) and one of (e,d), (f,d), or (g,d)

9. a) ∞, b) 6: if $\underline{X}$ = BCE, $\underline{Y}$ = BDFG, and $\underline{Z}$ = BCEFG, then AXXXYH,

A<u>XZZ</u>H, A<u>ZXZ</u>H, A<u>YYZ</u>H, A<u>YZY</u>H, A<u>ZYY</u>H.

10. a)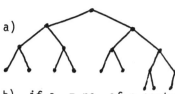
    bottom vertices = contestants, other vertices = matches

b) if a_n = no. of n-contestant tournaments, $a_n = \sum_1^{n-1} a_k a_{n-k}$,
$a_n = \frac{1}{(n+1)} \binom{2n-2}{n-1}$.

11. a) Build a "tree"-like graph as in Exer. 10a, where varieties are like contestants and intermediate stages are like matches,
b) 7·each split adds an additional variety eventually.

12. (a,c), (b,c), (c,d), (d,e)

13. a) bf, bg, fc, fd, ce, b) (a,b), (a,g); (c,d), (d,e); (d,e), (e,f), c) 3-remove edges incident to a given vertex

14. $\frac{28-5}{28}$: $\binom{8}{2}$ = 28 pairs of edges, of which 5 pairs disconnect

15. Block surveillance 6, e.g., b,c,e,g,h, j; corner surveillance 3: c,e,k or a,d,j

17. • block surveillance
    □ corner surveillance a) b)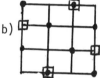

c)

18. a) 5, e.g., use squares with coordinates (2,4), (3,4), (4,4), (5,4) and (8,4), b) 10, e.g., (2,4), (2,5), (3,4), (3,5), (4,4), (4,5), (6,4), (6,5), (7,4), (7,5)

19. Jobs a,b,c collectively have only two qualified people

20. a) A-c, B-a, C-b, D-d or A-a, B-b, C-d, D-c or A-c, B-b, C-d, D-a

22. a) a-b, c-d, e-f, c) None, a,c,e collectively must be matched with b and d

26. a) 20, b) 10

29. a) yes, b) yes, c) no (contains Fig. 7d), d) no

30. b) [diagram], or [diagram], or [diagram], c) [diagram], and others.

33.

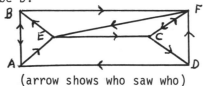

34. K_6 plus isolated vertex, yes

37. Need to find person whose information, if disregarded, makes the graph interval. Eliminate 4-circuits ABFD, AECD, AEFD, the liar must be D.

(arrow shows who saw who)

40. a) Let T and R be essentially equal, likewise for T' & R',
    b) A B has CB model

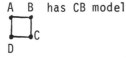 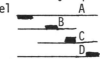

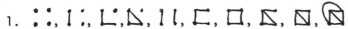

41. a)
```
    S
   /|\
  B A S
   /|\
  B A S
   /|\
  B A S
   /\
  B A
```

b) No way to get sequence beginning with double B's.

Section 1.2

1. (symbols)

2. (symbols)

3. No   4. No (only left graph has circuit of length 5).

5. a) No, degrees different, b) yes, a-5, b-6, c-2, d-1, e-4, f-3, c) no, look at complements, d) no, look at neighbors of the degree-5 vertex in each graph, e) no, only right graph has triangles, f) yes, a-2, b-3, c-5, d-7, e-6, f-8, g-1, h-4, g) no, only left graph has triangles, h) no, 3 overlapping triangles bac, acd, cde have no possible counterparts.

6. 1st ≈ 3rd ≈ 6th ≈ 7th and 2nd ≈ 4th ≈ 5th

7. 1st ≈ 2nd ≈ 3rd   8. No

9. a) No, only vertex 5 has 3 edges all inward

11. Both have size 3

13. a) Yes , b) No, odd number of vertices of odd degree,

c) Yes,

Section 1.3

1. a) 12, b) 9, c) 8 or 10 or 20 or 40

2. a) yes, b) yes, c) no, d) no

3. $2E \geq 3V$ or $\frac{2}{3}E = \frac{2}{3}(19) \geq V$, answer is 12.

4. n-1 must be even number; deg (x) + $\overline{\text{deg}}$ (x) = n-1 = even. So deg (x) is odd <=> $\overline{\text{deg}}$ (x) is odd. Thus $\overline{G}$ has same number of odd-degree vertices as G, namely, n-1.

5. pV = 2E => E = pV/2. 2 cannot divide p, so E = $p(\frac{V}{2})$ and E is a multiple of p.

6. Consider graph of games within one conference: 13 vertices, each of degree 11 - impossible.

Section 1.4

1. a) 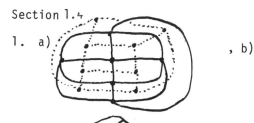 , b)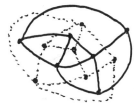

2.   no way to add

41

3. a) remove edges (a,f) & (c,d) to get $K_{3,3}$ config,
   b) remove (e,f) & (b,c) to get $K_{3,3}$,
   c) remove (b,j) & (e,g) to get $K_{3,3}$,

   d) remove vertex f, edges (c,h), (a,e), (a,f) to get $K_{3,3}$,
   e) remove (c,d), (d,e), (a,b), (f,g) to get $K_{3,3}$,
   f) remove (e,f) to get $K_{3,3}$,
   g) remove (b,c) to get $K_{3,3}$,
   h) remove (c,d) to get $K_{3,3}$.

5. No, messy but many $K_{3,3}$ configurations.

6. a) $L(K_5)$ has 10 vertices of degree 6, so $E = \frac{6 \cdot 10}{2} = 30 \nleq 3 \cdot 10 - 6$; $L(K_{3,3})$ contains a $K_{3,3}$.

   b) $L(H) = K_5$, where $H = $ 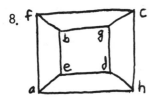 ; so $L(H)$ is non-planar.

7. Project plane onto sphere with North Pole deleted in 1-1 continuous fashion with origin $\rightarrow$ South Pole.

8.

9. By deleting edges and/or vertices, $K_{3,3}$, K_5, K_6 are obtained from K_7 on torus (identify lines x and lines y to convert rectangle into a torus)

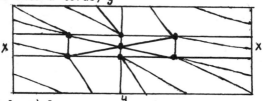

10. a) 1, b) 1, c) 1

11. a) i) yes, ii) yes, iii) no - removing f leaves non-planar graph

13. a) $R = E-V+C+1$,
    b) With part a), corollary is $E \le 3(V-C-1) \le 3V-6$

14. $2E = \Sigma \deg (x) \ge 3V$ or $V \le \frac{2}{3}E$. $V = E-R+2 \le \frac{2}{3}E$ or $E \le 3R-6$

16. a) If result false, then $2E = \Sigma \deg (x) \ge 6V$ or $3V \le E$, but $E \le 3V-6$
    b) Result true for each component.

17. $2E = \Sigma \deg (x) \ge 5V$. So $\frac{5}{2}V \le E \le 3V-6 \Rightarrow 5V \le 6V-12$ or $V \ge 12$, but $V < 12$.

19. $2E = \Sigma$ deg (region) $\geq rR$. So $\dfrac{2}{r}E \geq R = E-V+2$ or $E \leq (r/r-2)(V-2)$

21. a) $K_{3,3}$ has no triangles since it is bipartite,

    b) Letting r = 4 in Exercise 19, we have $9 = E \leq \dfrac{4}{2}(V-2) = 8$.

7. Supplement II

1. $\binom{7}{2} = 21$

2. a) Trivial by definition, b) all $\binom{n}{2}$ pairs joined by edges.

3. $3V = \Sigma$ deg $(x) = 2E = 2 \cdot 18 \Rightarrow n = 12$.

4. Pigeonhole Principle, same as Exercise 2 of section 1.4.

5. In $\overline{G}$, vertices in different component of G are joined by an edge. Two vertices from same component of G are linked by a path of length 2 whose middle vertex is a vertex in other component.

6. Start tracing a path until some vertex z is repeated (since deg ≥ 2, each vertex entered on path can be left on different edge). Between first and second visit to z a circuit is formed.

7. G-x is not connected if and only if there is no path without x in G between some 2 vertices.

8. If G_1 were a component of G with just one odd-degree vertex G_1 would violate Cor. in section 7.3.

9. 11 is the smallest integer n such that $\binom{n}{2} \geq 50$.

10. a) Yes - same argument as in Exercixe 6,

    b) No

11. In such a partition (V_1, V_2), there is no path from vertices in V_1 to vertices in V_2 and so G is not strongly connected. Thus if G is strongly connected, no such partition can exist. If G is not strongly connected, then let x be chosen so that x does not have a path to the set V_2 of vertices $(V_2 \neq 0)$, but does have paths to the other vertices V_1. Thus (V_1, V_2) is the desired partition.

12. a) Yes, see Exer. 7a of section 7.1,
    b) No, no way to have paths in both directions between a and b,
    c) Yes, many ways.

13. Condition is clearly necessary. Sufficiency (sketch): find a circuit in G and direct edges to make circuit strongly connected in initial graph G_1; successively add side path from vertex in G_i to new vertices and back to G_i.

30. a) A 5-circuit.

31. Pick any vertex, call it x. Assume, by symmetry, that a majority (at least 3) of edges incident at x are red. Let 3 of these red edges go to vertices a, b, c. If there is a red edge between two of a, b, c, then a red triangle (with x) results. Otherwise, a, b, c form a white triangle.

32. a) possible,   b) not possible.

Chapter II

Section 2.1

1. a) 1-2-3-4-5-6-7-1-4-7-3-6-2-5-1,
    b) c-a-e-b-a-h-g-b-c-d-g-f-e-d-f,

2. a) n odd,   b) yes, K_2

3. Possible, for example

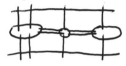

4. 2 times

5. Not possible, equivalent
    to finding Ham. circuit.

6. With hinted edge added, graph now has all even degrees and
    hence an Euler circuit. Removal of extra edge leaves Euler
    path.

7. a) Use same argument as in proof of Theorem,
    b) At each vertex, pair incoming with outgoing edges, and con-
        tinue as before,
    c) a-b-c-d-b-h-i-j-a-i-c-f-g-f-e-h-g-e-d-h-a.

8. A directed multigraph has an Euler path but not Euler circuit
    if and only if it has in-degree equal out-degree at all but
    2 vertices, at which in-and out-degrees differ by 1. Proof
    is same as Exercise 4's proof of Corollary.

9. The Euler circuit provides a path from any vertex to any
    other vertex.

10 a) A bridge cannot be on any circuit.

    b)

11. A circuit can theoretically have no edges, i.e., an empty
    sequence of edges. In a 1-vertex graph, such a   vertex is
    an Euler circuit.

12. △ .   This graph has "Euler circuit" but is not connected.

14. No, the 8 squares located at side of board one square from
     corners have odd degree.

15. Treat matrix as adjacency matrix of a graph. Desired sequence
     is simply an Euler path (which exists) in this graph: AB, BC,

CE, ED, DB, ӨF, FD, AE, EF.

16. a) Each edge has an odd number of adjacent edges at a vertex; so in all (both end vertices), it has odd and odd=even degree. Hence each vertex in L(G) has even degree, and L(G) is connected because G is.

   b) K_4.

Section 2.2

1. a) 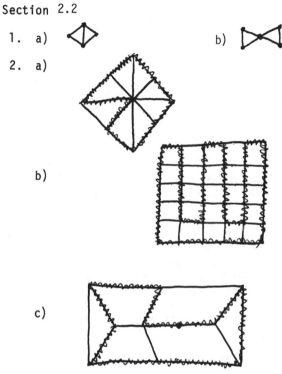   b)

2. a)

   b)

   c)

3. a-b-f-e-g-j-k-h-c-a

4. a) path: b-a-c-e-d; use rule 1 at b and c to force subcircuit.
   b) path: a-d-g-h-e-b-c-f-i; use rule 1 at d, e, f, by symmetry at b use edge (a,b)-now stuck at c.
   c) path: f-a-d-g-b-e-h-c; use rule 1 at c,e to force subcircuit.
   d) path: j-c-a-b-d-f-i-h-e-g; to reach a and b, must go - c-a-b-d- or c-b-a-d, similarly for i,f and g,e, but pasting these subpaths together leaves no way to visit j.
   e) path: j-l-g-h-i-f-e-d-a-b-c-k-m; same argument as in Example 3.

46

f) path: a-b-c-d-e-j-g-i-f-h; long case-by-case argument needed, start by symmetry with -a-f-i-.

g) path: b-f-i-e-d-a-c-g-j-k-h; long case-by-case argument, start with i) b-f-i, ii) b-f-c, iii) b-f-j.

5. a-b-c-d-m-p-o-l-k-j-i-q-e-f-g-h-n; by symmetry pick 2 edges at q, say (a,q) and (e,q) and use rule 1 at c, g, k. At b and d, one but not both use edge to m-by symmetry use (a,b) and (d,m) (delete edge (d,e)). So we have subcircuit p-m-d-c-b-a-q-e-f-g-h-n-p (h-n-p part is forced by deletion of (f,n)).

6. a) i) Apply rule 1 at j, f, h to form subcircuit,
   ii) messy, start at p and show middle path is

7. a) $2r_4 = 2r_4'$ but $r_4 + r_4' = 3$,

   b) $4r_6 + r_3 = 4r_6' + r_3'$ but $r_6 + r_6' = 3$, $r_3 + r_3' = 1$,

   c) $2r_4 = 2r_4'$, but $r_4 + r_4' = 9$.

8. a) A Hamiltonian circuit alternates red and blue vertices and so the number of each color must be equal. If a Hamiltonian path ends at same color at which it starts, numbers of red and blue could differ by 1.
   b) i) a,e,f,g,k one color, b,c,d,h,i,j other-colors not equal
      ii) a,c,e,h,j,l,m,n,o one color, b,d,f,g,i,k,p any color-colors not equal.

9. a) At most $E-E'$ edges can be used in Hamiltonian circuit, but a Hamiltonian has V edges.
   b) The sum for E' counts some edges twice if I is not a set of non-adjacent vertices and hence E' is not a correct bound on the number of edges that cannot be used.
   c) i) I = a,c,e,g,i,n,
      ii) I = a,e,f,g,k,
     iii) I = b,d,f,g,i,k,p

14. Number of students moving to left=number moving to right; so total moving left or right is even. Similarly total moving forward or backward is even. Hence total moving is even, but total=25.

Section 2.3

1. a) 3, has K_3,
   b) 4, use method in Ex. 3,
   c) 2, has K_2,
   d) 4, use method in Ex. 3,
   e) 3, 5-circuit cannot be 3-colored,
   f) 4, use method in Ex. 3 (hint: redraw with outer circuit a,d,g,c,f,b,e)

g) 4, to show 3 colors fail, show that in any 3-coloring of (a,b,c,d,e,f,g,h), a and h must have same color (as if connected by an edge (a,h)) and similarly for i and p; so that a,h,i,p effectively are a K_4.

2. a)

   b)

   c) requires 4 edge colors

3. a) Graphs in part b,f,g,   b): a)

   c)   d) remove any 1 vertex,

   e)

4. a) bfg, ceh, aeg, aeh   b) bdf

5. a) abc, def, gh,   b) abe, cd, fg,

   c) abc, def, ghi, jk,   d) ai, bcd, efg, hp, jkl, mno.

6. a) Equitable   b) Equitable

7. a) $k^4-4k^3+6k^2-3k$,   b) $k^5-5k^4+9k^3-7k^2+2k$

   c) $k(k-1)(k-2)(k-3)$

   d) Find smallest positive integer k for which $P_G(k)$ is positive.

8. No, Nevada (or Kentucky or West Virginia) and its neighbors form an odd-length wheel (see Example 3).

9. Assume n-1-intersecting-circles picture can be 2-colored. Then draw in n-th circle and complement colors inside new circle.

10. 3 colors (coloring is straightforward).

12. Vertex=banquet, Edges join banquet vertices with common room, color=day of week. Can graph be 7-colored?

13. a) Vertex=experiment, edge=overlap of experiments, color= observations.

   b)   requires 4 colors (similar to Example 4).

48

1. a) At most n-d vertices of one color (a given vertex and all
   its non-neighbors); so $(n-d)\chi(G) \geq n$,
   b) $K_{3,3}$ (or any $K_{n,n}$).

2. a) If not connected, one component would be color critical,
   b) A vertex x of deg < k-1 can always be properly k-colored
   after G-x is k-colored,
   c) If x is articulation point, let C_i be components of
   G-x; color each $C_i \cup x$ with x same color and paste together.

3. a) The k-chromatic critical subgraph is obtained by successively deleting a non-critical vertex (whose removal does not
   reduce $\gamma(G)$) until all remaining vertices are critical.
   b) If a graph G cannot be k-colored, then G or some subgraph
   is k+1-chromatic and by part a has a critical subgraph.
   c) It suffices to show that no connected graph without odd
   circuits has a 3-chromatic critical subgraph. If G is such a
   subgraph, then first 2-color G-x, for some x. Since G cannot
   be properly 2-colored, x must have neighbors a and b of different colors and there must be a 1-2 path between a and b
   (or else a's color could be changed as in the proof of Theorem
   5). But this 1-2 path plus x forms an odd circuit.

4. If y and z were adjacent and paths P_y and P_z from x to y and z,
   resp., are both even (or both odd), then $P_y,(y,z),P_z$ forms an
   odd circuit (or contains an odd circuit if P_y and P_z have common vertices or edges).

5. Two edges are adjacent in G if and only if the associated vertices are adjacent in L(G).

6. a) A maximal set of vertices of one color in G, say the set's
   size is k, corresponds to a complete subgraph in $\overline{G}$. Hence
   $k \leq \chi(\overline{G})$. Then $n \leq \chi(G)k \leq \chi(G)\chi(\overline{G})$.
   b) Using induction, assume $\chi(G-x)=k$ and $\chi(\overline{G}-x) \leq n-k$ (G has n
   vertices). If deg(x)<k, then in G give x one of the remaining
   k-deg(x) colors (different from x's neighbors' colors) and in $\overline{G}$
   use a new color for x; so $\chi(G)+\chi(\overline{G}) \leq k+(n-k+1)=n+1$. If deg(x)$\geq$k
   and $\overline{\deg}(x)<n-k$, interchange roles of G and $\overline{G}$ (recall deg(x)+
   $\overline{\deg}(x)=n-1$).   c) Follows from b.

7. Prove by induction that all circuits are even and then use
   Theorem 4. Sketch of proof: induct on number of regions inside circuit. Trivial if one region inside. If several regions inside circuit, break into one region and rest of regions.
   Show if circuit odd then one of the 2 subareas must have odd-
   circuit boundary.

8. By Euler's formula, R=E-V+2=13-8+2=7. If minimal circuit is $\geq$4,
   then 2E=Σdeg(region)$\geq$4r or E$\geq$2r=2·7=14, but E=13. Hence G has
   a triangular region whose vertices require 3 colors.

Chapter III

Section 3.1

1. a) 2: ⊥ , ⊔

   b) 3: ⊻ , ⊔ , ∧

   c) 6: ⊻ , ⊻ , H , ⊔⊔ , ⊥∧ , W

2. 21 (such a graph must be a tree).

3. They have no odd-length circuits. So by Theorem 4 of section 8.4, trees are 2-colorable.

4. They have no K_5 or $K_{3,3}$ configurations (which involve circuits).

5. a) No circuits => unique path from root to any vertex,
   b) Some subset of edges in a connected graph forms a tree containing all n vertices. This tree has n-1 edges (by Theorem 1) and so must contain all the edges.
   c) This condition implies no circuits. Now see part a.

6. Trees, in a planar depiction, have one (infinite) region; i.e., $1=R=E-V+2$ or $E=V-1$.

7. Vertices of degree 1 are leaves or possibly a root: if root of degree 1, pick root and a vertex (leaf) at largest level. If root x of degree ≥ 2, pick vertex (leaf) at largest level in each of 2 subtrees rooted at sons of x. Alternate proof: starting at any vertex, trace a path until path ends; i.e., one comes to vertex x of degree 1; next repeat process starting from x.

8. a) $\ell+i=n$ & $n=mi+1 => \ell+i=mi+1$ or $\ell=(m-1)i+1$,
   b) solve for i in part a, $i=(\ell-1)/(m-1)$ & $n=i+\ell =>n=(\ell-1)/m-1 + \ell=(m\ell-1)/(m-1)$,
   c) solve for i in Theorem, $i=(n-1)/m$ and for ℓ in part b, $\ell= [(m-1)n+1]/n$.

9. Starting at the root and moving down the tree (to larger levels) with m choices at each corner (internal vertex), there are at most m^h different paths of length h.

10. $(m^{h+1}-1)/m-1=m^h+m^{h-1}+...m+1$.

11. By Corollary part c, $i=(n-1)/m$, fraction is $(n-1/m)/n=\frac{1}{m}(\frac{n-1}{n})\approx\frac{1}{m}$.

12. n-t

13. If tree is balanced with all leaves at 1 or 2 different levels, then result is immediate from Theorem 3. If a balanced tree is "unbalanced" by deleting 2 leaves (letting their father become a leaf) and making them sons of some leaf at the same or greater

50

level, then the average leaf level increases. Any binary tree can be obtained from a balanced tree by a sequence of such alterations.

14. If tree is rooted and fathers colored before sons, then each vertex, except root, will be adjacent to only one previously colored vertex. If k colors available, root has k choices, other vertices k-1 choices. In total, $k(k-1)^{n-1}$ choices.

17. a) Internal vertices = arithmetic operations,
    b) 7

20. 8 losers tournaments of sizes 24, 16, 12, 8, 6, 4, 3, 2

22. Check if X=A, then if X=B, if X=C, etc.

25. a) Name coins 1, 2, 3, 4, T

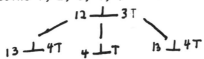

    b) If first weighing is 1|2 and balances, then 4 possibilities are left (only 3 can be distinguished in one weighing). If first 12|34 and scales tip right, 4 possibilities are left.

Section 3.2

1. a) Several easy depth-first search paths from S to E
   b) 4

2. Breadth-first search path is simply shortest path from S to E, easily seen by inspection.

3. (0,0)-(7,3)-(0,3)-(0,4)-(6,4)-(6,0)-(2,4) [order of choice at each vertex: left, right, up, down, diagonal]

4. a) (0,0)-(5,0)-(2,3)-(2,0)-(0,2)-(5,2)-(4,3),
   b) (0,0)-(0,5)-(7,5).

5. (0,0)-(0,4)-(4,0)-(4,4), now 2 quarts left in 10-quart pitcher.

6. Several possible solutions: one way continues search in Example 2 (2,4)-(2,0)-(0,2)-(7,2)-(5,4)-(5,0)-(1,4).

10. Graph has 22 vertices.

11. a) See Busacker & Saaty, Finite Graphs and Networks (McGraw-Hill), p. 156-157 for this and related problems.

12. b) 14 (along 2 opposite edges of the board)

14. 5 (2,4), (3,4), (4,4), (5,4) & (8,4)

Section 3.3

1. 11:1-3-2-4-1

2. 20:1-4-2-6-3-5-1

4. Min tour is 1-4-3-2-5-6-1 of cost 18. An "ideal" tour, call it S, would use the 2 cheapest edges (one entering and one leaving) at each vertex; i.e., 2 cheapest entries in each row. See underlined entries in matrix. This ideal tour S costs 1/2 (sum of underlined entries)= 1/2(33)=16 1/2 [divide by 2 since each entry would be counted twice]. However, such an S contains (at least) 3 entries in columns 1 and 3. So a feasible tour would have to use at least 2 second-choice (costlier) entries. Then a feasible tour must cost at least 1/2(33+2)=17 1/2; i.e., at least 18. Thus the given tour is minimal.

$$\begin{array}{cccccc} \infty & 3 & 3 & \underline{2} & 7 & \underline{3} \\ \underline{3} & \infty & \underline{3} & 4 & 5 & 5 \\ 3 & 3 & \infty & \underline{1} & 4 & 4 \\ \underline{2} & 4 & \underline{1} & \infty & 5 & 5 \\ 7 & 5 & \underline{4} & 5 & \infty & \underline{4} \\ 3 & 5 & 4 & 5 & \underline{4} & \infty \end{array}$$

5. $T_1-T_5-T_4-T_3-T_2-T_1$

7. a) 1-2-4-3-1,
   b) 1-5-3-4-6-2-1,
   c) 2-3-5-1-4-2

Section 3.4

1. a) Tree will be a simple path.
   b) a-b-c-d-e-f-g-h,
   c) a-e-b-f-c-g-d-h

2. a) 7 leaves at level 1,
   b) a is root, others are sons of a, except d and f,

   c)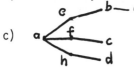

4. Depth first spanning tree is path $x_1-x_3-x_5-x_7-x_4-x_2-x_6$

5. 4 components, x_{17}, x_{19}, x_{23} are isolated vertices, other component is the breath-first search tree $x_2-x_4-x_6-x_3-x_9-x_{12}-x_8-x_{10}-$
$x_5-x_{15}-x_{20}-x_{14}-x_{25}$ $x_{16}-x_{18}-x_{22}-x_{24}-x_{26}-x_{13}$ x_7-x_{21} 11

6. a) a,b,d,e,h,ℓ,i,f,c,g,j,k and d,ℓ,h,i,e,f,b,j,k,g,c,a

7.

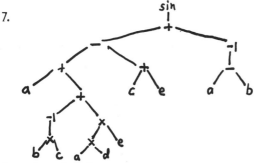

8. a) A(,) is adjacency matrix, n=|V|, and Pre( ) is pre-
   order index list of vertices (initially all 0's):
   Pre(1)←1 ;   Call Tree (1,1);
   Tree (j,i); for k←1 to n do if A(j,k)=1 & Pre(j)=0 then
      begin i←i+1; Pre(k)←i; Call Tree (k,i) end;
   return.

9. If such a graph were not a tree, it would have a circuit - involving tree edges and one non-tree edge. Replacing a tree
   edge in the circuit by the non-tree edge would yield a new tree.

11. a) Breadth-first search builds shortest paths from a to each
    vertex. This minimizes longest path (level).

    b) Graph a b c d . Breadth-first at a has height 2,
    while depth-first at b has height 1.

14. a) distinct vertices in diff. biconnected comp. are not
    adjacent,

    b) let there be 2 biconnected comp. C_1,C_2 joined at vertex a.
    If C_1,C_2 planar, choose planar depictions of C_1,C_2 each
    with a on infinite region. Join at a.

    c) same reason as part a.

23. n-1

24. In a postorder traversal, sons are checked before their
    fathers.

27. Since a depth-first spanning tree has no cross edges, subtrees
    rooted at sons of x (the root) are joined only at x.

32. a) Since a preorder traversal visits each internal vertex be-
    fore its (2 or more) descendants, a characteristic sequence
    must have at least two 0's after the last 1.

Section 3.5

2. After the k-1 smallest items, the k-th smallest item is first

(smallest) remaining item on one of the 2 sublists. So one comparison is needed to find k-th smallest item, for k=1,2,..., n-1. The n-th item is the one left over.

3. A binary comparison tree has n! leaves (at least one sequence of comparisons for each possible rearrangement). So average number of comparisons = average leaf level = $\log_2 n!$ = $O(n\log_2 n)$.

4. a) Vertex = each sublist formed. A sublist is split into a first and second sublist (left and right sons).

   b)

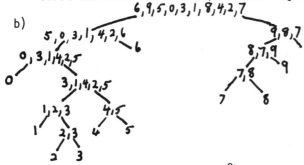

5. a) List 1,2,3,4,5,6,7,8 needs $\binom{8}{2}$ comparisons. In general, 1,2,...,n.

   b) Let the first item in the list be 1 or n, and for each successive item pick either the largest or smallest unused number. 2^{n-1} such lists. This way only one item will be in one sublist at each stage.

6. let $a_0 = a_1 = 1$, $a_n = (n-1) + \dfrac{1}{n} \sum_{i=1}^{n} (a_i + a_{n-i}) = n-1 + \dfrac{2}{n} \sum_{i=0}^{n-1} a_i$.

   Can show by substitution that $a_n \leq 2n\log_2 n$

7. a) Assume a balanced heap of first k elements has been made (k=1, trivial). Attach k+1-st element, a new leaf, as either the son of a 1-son internal vertex or else the son of a former leaf at the largest level. Now repeatedly compare new element with its father and interchange the 2 elements if father is smaller.

   b) 1 part a)

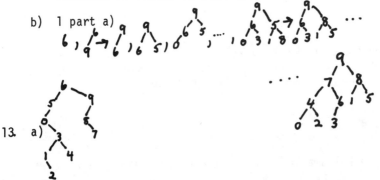

13. a)

13. b) The tree in part a and the tree in Exercise 4b are the same. The first element in each sublist in 4b is the element at the corresponding vertex in part a's tree.

c) In general, "tree sort" adds each successive element in the list being sorted to the tree by a sequence of comparisons with earlier elements in the list - these comparisons correspond exactly to the comparisons of that element with earlier elements (at the front of sublists) in Quik sort.

14. b) $O(n\log_2 n)$

Chapter IV

Section 4.1

1. 14:c-d-h-k-j-m

2. a) 19:a-f-g-ℓ-q-v-w-x-y,
    b) 21:d-c-h-m-r,
    c) 15:e-j-i-h-g

3. a) 31:L-c-d-f-g-k-W,
    b) 32:L-b-h-j-m-W,
    c) 13:L-a-c-d-f-g-k-W,
    d) L-c-d-f-g-k-w,2

4. a) L-c-d-f-g-k-W (R=57),
    b) L-c-d-f-i-k-W (R=63)

5. a) 5:L-b-h-j-m-W,
    b) 6:L-c-d-f-g-k-W,
    c) 6:L-c-e-g-j-m-W

6. Same as answer in 3a.

11. If edges used in labeling are "directed" towards vertex being
    labeled, then each vertex has one incoming edge, except for
    starting vertex (root). This is exactly the structure of a
    rooted spanning tree.

Section 4.2

1. a) 39:(N,b), (b,c), (c,d), (d,h), (d,e), (e,f), (e,g), (f,i),
        (g,j), (j,k), (j,m), (m,R),
    b) same as a.

2. a) 59:(L,a), (a,c), (c,d), (b,d), (d,e), (d,h), (h,f), (f,g),
        (g,j), (g,i), (i,k), (k,ℓ), (k,m), (k,W)
    b) 60:replace (k,W) by (m,W) in part a,
    c) several possibilities, e.g., L,j cheapest path is 17, but
        tree path length is 22.
    d. 24:similar to part a but delete (a,b), (g,i), (k,W) and
        add (a,b), (g,k), (m,W).

3. Does not alter answer in 2a.

4. Cost is 63.

5. 133, just apply algorithm
    with "shortest" replaced
    by "longest".

6. Make a simple path of 8 vertices and 7 edges, each edge of of cost 1. Add any 8 more edges with cost 100.

9. Let T be minimal Prim spanning tree and let T' be any other minimal tree. The proof of Prim's algorithm now does not admit the possibility that e_k's length equals e_j's length; so depending on whether e_k or e_j is shorter, either T' or T is not minimal-contradiction.

Section 4.3

1.

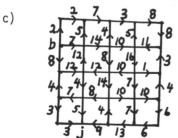

2. a) Max flow = 21, P = {a,b,c,d}
   b) Max flow = 24, P = {a,b,c},
   c) Max flow = 14, P = {a}

3.

4. a), b) Max flow of 19 saturating all edges into sink is easily found many ways.
   c) A flow of 15 can saturate $\{P,\bar{P}\}$, P = {L,a,b,h,j} of cap. 20.
   d) Impossible: more flow must go into h than can leave h.

5. a) Max flow - 13, P = {a,b,c}

   c)

6. Yes. Build flow paths by choosing uppermost unsaturated edge from each vertex.

57

7. Set capacities of edges from a and to z to 100. Max flow = 150, $P = \{a,b,c,d\}$.

8. a) Not possible $k(P,\overline{P}) = 50$, $P = \{a,b,c,d,e,g\}$.
   b) Now possible.

9. 5 messengers

11. a) 3
    b) 9, give edges capacity 3, min cut has $P = \{L\}$.

12. Split b,c,d into 2 vertices, one for incoming edges and one for outgoing, join the 2 vertices by an edge of capacity 5.

    E.g., d becomes:

    Max flow = 9, $P = \{a,b_1,b_2,c_1,c_2,d_1,e_1,d_2,e_2,f_1\}$

13. See Exercise 12. Max flow = 40, $\overline{P} = \{g_2,f_1,f_2,z\}$

14. 3

15. Max flow = 2100, route flow as follows: 4 on $a_0-d_2-z_3-z_4$, 3 on $a_0-c_2-z_4$, 4 on $a_0-b_1-z_3-z_4$, 2 on $a_0-b_1-c_2-d_3-z_4$, 4 on $a_0-a_1-d_3-z_4$, 4 on $a_0-a_1-b_2-z_4$. $P = \{a_0-a_4,b_2,b_3,b4,c_3,c_4,d_4\}$

22. a) Give each vertex capacity 1 (see Exer. 12) and each edge capacity ∞. Find max flow between each pair of vertices. If max flow is 1 for some pair of vertices, min cut is articulation point. If max flow is always ≥ 2, then 2 of the unit-flow paths between any given 2 vertices form a circuit.
    b) Give each edge capacity 1. If max flow between each pair of vertices is k (k disjoint paths) this is equivalent to the min cut having capacity $\geq k$, i.e., G is k-edge-connected.

25. a), b) Assume network undirected. Convert to multigraph, with the number of edges from x to y equal to $\phi((x,\overrightarrow{y}))$. Now repeatedly apply Euler path building method from a (always ending at z), thus building $|\phi|$ a-z paths.

39. See Ford and Fulkerson, <u>Flows in Networks</u>, (Princeton Univ. Press, 1962) p. 21.

Section 4.4

1. a) A-Bi,D-La,Lo-G,Bo-F,C-J is one possibility.

   b) Using other pairs of A-G and Lo-J, the labels in the algorithm are: $a(-,\infty),Bo(a^+,1),F(Bo^+,1)$, $G(Bo^+,1),C(F^-,1)$,

58

1. b) $A(G^-,1),J(C^+,1),Bi(A^+,1),Lo(J^-,1),D(Bi^-,1),La(Lo^+,1),z(La^+,1)$.

   New matching is A-G,D-Bi,Lo-La,Bo-F,C-J.

c)
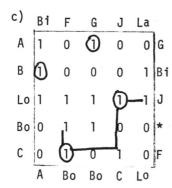

2. c)
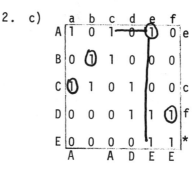

2. a) A-a,B-b,C-d,D-e,E-f is one possibility.

   b) Labels are $a^*(-,\infty),E(a^{*+},1),e(E^+,1).$ $f(E^+,1),A(e^-,1),$
   $D(f^-,1),a(A^+,1),c(A^+,1),d(D^+,1),C(a^-,1),z(a^+,1).$

3. Several solutions, see Exercise 2a or 2b.

4. Give edges from a and to z the appropriate supplies and de-
   mands. Middle edges still ∞. Possible pairing Bi-A(3 dates),
   Bi-D(2),F-Bo(1),F-Lo(2),F-C(1),G-A(1),G-Bo(2),J-C(3),J-Lo(2),
   La-D(3).

5. Give edges from a and to z cap. 2 and middle edges now cap. 1.
   Possible solution: Bi-A,Bi-D,F-Bo,F-C,G-A,G-Bo,J-C,J-Lo,La-D,
   La-Lo.

6. No, the schools with demand of 7 Ph.D.'s can only hire at most
   6 (1 from each university).

7. Jobs 2,3,4,5,6.

8. Max is 3, possible assignment (1,2),(2,4),(3,6),(4,5),(5,3).

9. Min is 3, possible assignment (1,4),(2,2),(3,5),(4,6),(5,3).
   (Just reverse roles of max and min.)

10. If less than n independent 1's, then by Theorem 1, less than
    n lines cover all 1's. Now $k(k\geq1)$ of the lines must be
    columns. These k column lines cover a number of 1's at least
    equal to the number of 1's in the >k non-line rows. This con-
    tradicts the fact that row sums exceed column sums.

12. a) None

12. b) 18 2 0 0 0 0
          0 16 18 6 0 0
          0 0 0 10 0 0
          0 0 0 2 11 0
          0 0 0 0 7 18

21. Theorem 2's condition applies to submatrices RxC' and R'xC
     (if it did not apply to RxC' and for A⊂R,|A|>|R(A)|, then a
     smaller 1's cover could be found by replacing A with R(A),
     and so RxC and R'xC have independent sets of 1's of sizes
     |R| and |C|, resp., or an independent set of size |R|+|C| for
     the whole matrix, as required.

Chapter V

1. a) 12, 56, b) 3, 15   2. a) 51 , b) 26, c) 46
                                                   100   30
3. $8 \cdot 10 \cdot 6 \cdot 4 = 1920$ 4. 1296   5. $2^{100} \approx 10^{30}$
6. $26^4 = 456, 976,$ 26!/22! = 358,800   7. $n(n-1)$
8. a) 20   b) $8 \cdot 7 \cdot 5 = 280,$ c) $3 \cdot 8 \cdot 7(7+5) + 3 \cdot 7 \cdot 6(8+5) + 3 \cdot 5 \cdot 4(8+7) = 4554$
9 a) 12,   b) 14 ,   c) $14^2 = 196,$ d) 196-4= 192 , e)122
10) a) $4 \cdot 47 = 188,$ b) 2 cases: first card is spade queen or is not:
                                $1 \cdot 48 + 12 \cdot 47 = 612$
11. (3+1)(5+1)-1= 23   12. Use induction :$1/2(6^{n-1})$ of
                              outcomes on first n-1 dice are even and,
                              composed with 3 even outcome on n-th
                              die, yields $1/2 \cdot 6^{n-1} \cdot 3$ even outcome.
Similarly first n-1 odd and n-th odd yields $1/2 \cdot 6^{n-1} \cdot 3$ even outcomes.
In total = $1/2(6^{n-1} \cdot 3)+ 1/2(6^{n-1} \cdot 3) = 1/2(6^n)$

13. 12   14. $5 \cdot 26 \cdot 26 \cdot 5, 26 \cdot 21 \cdot 21 \cdot 26$
15. a) $9 \cdot 10^4$, b) $9 \cdot 5 \cdot 10^3$, c) $9 \cdot 9^3 + 4 \cdot 8 \cdot 9^3 = 29,889,$ d) $9 \cdot 10^2$
18. a)$6^4 = 1296$, b) $6^2/6^4 = 1/64,$
     c) $1/64 (1^2+2^2+3^2+4^2+5^2+6^2+5^2+4^2+3^2+2^2+1^2)= 146/1296$
19. a) $10 \cdot 9 \cdot 8^2 = 5760,$ b) 1/8   20. a) $2 \cdot 26^3 \cdot 10^3$
    20. b) $(2 \cdot 10^1 + 3 \cdot 10^2+4 \cdot 10^3)(26^1+26^2+26^3)$
21. $6 \cdot 5 \cdot 4 - 5 \cdot 4 \cdot 3 = 60$
23. (Seq. with e) + (Seq. with f but not e)= 60 + 36 = 96
24. Place one "9", then one "8", then non-(8,9) in 2 remaining positions
                              $4 \cdot 3 \cdot 8^2/10^4$
30. Every other even number is divisible by 4: $5 \cdot 5 \cdot 5 \cdot 1 = 125$
35. (1+1)(2+1)(2+1)(1+1) (3+1) = 144
40. There are 44 - 2i ways to place second Queen if first Queen is
 i squares from edge of board, for i=1,2,3,4. Answer is
          $1/2 (28 \cdot 42+20 \cdot 40+12 \cdot 38+4 \cdot 36)= 1288$

Section 5.2

1. 52! 2. $P(26,10) = 26!/16!$

3. $\binom{15}{9}9! = 15!/6!$ 4. $9!, \quad 4! \cdot 5!$ 5. $8!/3!$

6. $\binom{26}{4}$ 7. $\binom{10}{5}, \binom{8}{3}$ 8. $\binom{15}{5}, \quad \binom{10}{3}\binom{5}{2} / \binom{15}{5}$

9. $(\binom{10}{8}+\binom{10}{9}+\binom{10}{10})/2^{10}$

10. a) $\binom{4}{4} \cdot 48/\binom{52}{5}$, b) $13\binom{4}{4} \cdot 48/\binom{52}{5}$, f) $\binom{13}{5} \cdot 45/\binom{52}{5}$,

 d) $13\binom{4}{3} \cdot 12\binom{4}{2}/\binom{52}{5}$, e) $10 \cdot 4^5/\binom{52}{5}$ (Ace high or low),

 c) $\binom{13}{2} \cdot \binom{4}{2}^2 44/\binom{52}{5}$

11. $\binom{4}{2}[\binom{6}{4}+\binom{6}{5}+\binom{6}{6}]+\binom{4}{3}\binom{6}{6}=136$, b) $[\binom{4}{2}\binom{6}{2}-\binom{3}{1}\binom{5}{1}]+[\binom{4}{1}\binom{6}{3}-\binom{5}{2}]+\binom{6}{4}=160$

12. a) $(8!)^3$, b) $(3 \cdot 7!)^3/8!)^3$,

 c) Use inclusion-exclusion: $\frac{1}{(8!)^3}[8(3 \cdot 7!)^3-\binom{8}{2}(3 \cdot 2 \cdot 6!)^3+\binom{8}{3}(3! \cdot 5!)^3]$

13. $3!6!8!5!$ 14. $(4!)^{13}/[\binom{52}{13}\binom{39}{13}\binom{26}{13}\binom{13}{13}]$

15. $\binom{10}{5} \cdot 5! P(21,5)$, count patterns of 5 vowels and 5 consonants: $6/\binom{10}{5}$

16. a) $2 \cdot 5!/6! = 1/3$ (b either before or after a), b) 1/2 (obvious by symmetry)

17. $\binom{20}{10}10!$ 18. $\binom{3}{2} \cdot \binom{5}{2}$ (counts the team with 2 women).

19. $n=12, (\binom{12}{4}=495)$ 20. $\binom{30}{5}^4 \cdot \binom{31}{5}^7 \cdot \binom{28}{5}/\binom{365}{60}$

23. a) $(3!8!/2!2!)/(10!/2!2!) = 1/15$ 27. $P(7,4) \cdot 1$ (position consonants first)

29. a) i) Code colors are one of $\{R,W,Y,G\}$ plus Bu & Bk: (which of $\{R,W,Y,G\}$: 4 choices) . (Bu or Bk in position of color earning white key: 2 choices) $\cdot$ (possibilities for other 3 positions, not all Bu & Bk: $3^3 - 2^3$ choices) = 152

36 $P(\binom{5}{3},3)$, $P(\binom{n}{3},3)$ 38. a) $2^{11} = 2048$, b) 2^{11}

46 1- Prob.(no pairs of birthdays) $=1- P(365,25)/365^{25} \cong 1 - .431 = .569$

53 a) (pick a pair: $13\binom{4}{2}$ choices) $\cdot$ (pick 4 other values: $\binom{12}{4}$ choices) (pick suit for each of the 4 other values: 4^4 choices) =
$$13 \cdot \binom{4}{2} \cdot \binom{12}{4} \cdot 4^4/\binom{52}{6}$$

54 2^n (pick any subset of distinct objects and then add identical objects, as needed

60. d) Let d_i be triangles with i corners being vertices of n-gon. Then $d_0 = \binom{n}{6}$, $d_1 = 5\binom{n}{5}$, $d_2 = 4\binom{n}{4}$, $d_3 = \binom{n}{3}$. Hint: for d_i, consider the subset of 6-i endpoints (vertices of the n-gon) of the lines that form the triangle.

Section 5.3

1. $6!/(!2!3!) = 60$ 2. $11!/(1!4!4!2!) = 34,650$
   3. a) $3^6 = 6561$,   b) $90/6561$

4. $3(6!/3!3!) + 3!(6!/3!2!1!) + 6!/2!2!2!$

5. a) $10![1/6!2!2! + 2/4!4!2! + 2/5!3!2! + 1/4!3!3!]$,   b) $\binom{2+3-1}{2} = 6$

6. $\binom{10+4-1}{10} = 286$ 7. $\binom{3+6-1}{3} = 56$ 8. $\binom{9+3-1}{9} - 3\binom{3+3-1}{3} = 25$

9. a) $\binom{10+3-1}{10} = 66$,   b) $\binom{15+3-1}{15} - \binom{4+3-1}{4} = 121$ (outcomes with ≥ 11 dimes),
   c) 11

10. $\binom{10+3-1}{10} + \binom{3}{1}\binom{9+3-1}{9} + \binom{3}{2}\binom{8+3-1}{8} + \binom{3}{3}\binom{7+3-1}{7} = 402$

11. $6!/3!2!1! + 6!/2!2!1!1! = 240$ (1st digit = 4), (1st digit =6)

12. $\binom{3}{1}5!/2!1!1!1! + \binom{2}{1}\binom{3}{1}5!/4!1! + \binom{2}{1}\binom{3}{2}5!/3!1!1! +$

   $\binom{2}{1}\binom{2}{1}5!/3!2! + \binom{3}{2}\binom{2}{1}5!/2!2!1! = 550$

13. a) $\binom{4}{2}5^2 = 120$, b) 2,2- distribution: $\binom{6}{2}\binom{4}{2}$ codes) + 3,1-dist.: $6 \cdot 5\binom{4}{3}$ codes)

    + (4 dist.: 6 codes) = 216

14. $\binom{r+n-1}{r}$ 15. $\binom{10}{6}[\binom{6}{2}8!/2!2! + 6 \cdot 8!/3!]$

17. $\binom{8+25-1}{8}$, $(1/\binom{8+25-1}{8}) \sum_{k=0}^{4} \binom{8-2k+12-1}{8-2k}\binom{2k+13-1}{2k})$

19. $1/2[\binom{3n+3-1}{3n} - 3\binom{(3n-2n-1)+3-1}{(3n-2n-1)}) + 1]$ (outcome of $n\alpha$'s, $n\beta$'s, $n\gamma$'s in each half is only counted once)

22. a) $\binom{15}{4} \cdot \binom{10}{5}$ (ways to position 2's)(one 0 followed by any arrangement of 5 0's, 5 1's)

27. (locate 7 non-S type positions in 5 "boxes" before, between, and after the 4 S's with at least 1 non-S position between each pair of S's: $\binom{4+5-1}{4}$ ways). (arrange the 7 non-S letters in the 7 non-S positions: $7!/4!2!1!$ ways) = 7350

29. $1/(3\,(\binom{4+4-1}{4})6!/2!2![3!2\binom{3+4-1}{3})\frac{5!}{2!2!1!} + 3 \cdot 2\binom{4+3-1}{4})\frac{5!}{2!2!1!}]$

63

Section 5.4

1. a) $\binom{40+4-1}{40}$, b) 1 $\qquad$ c) $\binom{36+4-1}{36}$

2. a) 5^{20} $\qquad$ b) $\binom{5}{2}20!/7!^2 2!^3$, c) $20!/4!^5$

3. a) $\binom{13}{4}\binom{13}{3}^3/\binom{52}{13}$, b) $(13!/5!5!2!)(39!/8!8!11!12!)/(52!/13!^4)$

   c) $4\ \binom{48}{9}/\binom{52}{13}$

4. a) $\binom{5+3-1}{5}\binom{6+3-1}{6}\binom{6+3-1}{6}\binom{4+3-1}{4} = 246{,}960,$

   b) $\binom{5+3-1}{5}\binom{6+3-1}{6}\binom{(6-3)+3-1}{(6-3)}\binom{4+3-1}{4} = 88{,}200$

5. $\binom{(18-8)+4-1}{(18-8)}\binom{(12-8)+4-1}{(12-8)}\binom{(14-8)+4-1}{(14-8)}$

6. (order 2 a's & 2 b's: $\binom{4}{2}$ways)· (space C's among a's & b's, as in
   Example 5: $\binom{(8-5)+5-1}{(8-5)}$ ways) $= 210$

7. $5!\ \binom{(5-2)+4-1}{(5-2)}$ $\qquad$ 8. $21!5!\binom{(21-4)+6-1}{(21-4)}$

9. a) $\binom{28+5-1}{28}$, b) $\binom{(28-5)+5-1}{(28-5)}$, c) $\binom{(28-20)+5-1}{(28-20)}$

10. $\binom{15+3-1}{15}$ $\qquad$ 11. Let $x_5 = 100-x_1-x_2-x_3-x_4 > 0; \binom{(100-5)+5-1}{(100-5)}.$

12. a) $P(n,k)$ $\qquad$ b) $\binom{n}{k}$

13. a) $4^3\binom{9+4-1}{9}$ $\qquad$ b) $P(4,3)\binom{9+4-1}{9},$

    c) 4^3 distribute teddies & fill in with lollipops

14. a) $n!/n^n$ b)$(1/n^n)$ (which box empty: n choices) (pick a box and
    given it 2 object:$(n-1)\binom{n}{2})\cdot(n-2)! = \dfrac{n(n-1)n!}{2n^n}$
    c)$(1/n^n)\cdot\binom{n}{2}[\binom{n-2}{2}\binom{n}{2}\binom{n-2}{2}(n-4)!+(n-2)\binom{n}{3}(n-3)!]$
    (2 boxes with 2 obj.), (1 box with 3 obj.)

15. a)$\displaystyle\sum_{k=0}^{4} \binom{k+2-1}{k}\binom{(8-k)+4-1}{(8-k)}$, b) $\displaystyle\sum_{k=0}^{4} \binom{8}{k} 2^k 4^{8-k}$

16. $\displaystyle\sum_{i=0}^{5}\binom{15-3i+4-1}{15-3i}$ 17.a) $\displaystyle\sum_{k=0}^{7} (36-5k)$, b) $\displaystyle\sum_{k=0}^{7} (36-5k)+ \sum_{k=0}^{2} (11-5k)$

18. a) distributions of 8 distrinct objects into 3 distinct boxes
    c) distributions of 10 district objects into 4 distinct boxes with
       same no. in boxes 1 & 2.
19. a) arrangement of n objects chosen with repetition from n distinct types.
    b) arrangements of 15 objects, 3 of each of 5 different types.

20. a) distributions of 6 identical objects into 31 distinct boxes;
$x_1+..+x_{31}=6$, $x_i \geq 0$

b) distributions of 5 identical objects into 3 distinct boxes with at
most 4 objects in box 2 and at most 2 objects in box 3;
$x_1+x_2+x_3=5$, $0 \leq x_i$, $x_2 \leq 4$, $x_3 \leq 2$

21. a) selections with repetition of 30 objects from 5 piles;
$x_1+\cdots+x_5=30$, $x_i \geq 0$

b) selections with repetition of 18 objects from 6 piles, at least
2 per pile; $x_i+\ldots x_6=18$, $x_i \geq 2$

23. a) $\binom{(40-10)+5-1}{(40-10)}$, b) $5[\binom{(40-8)+4-1}{(40-8)} + \binom{(39-8)+4-1}{(39-8)}]$,

c) $\binom{(40+5-1}{40}$ $-5\binom{(40-21)+5-1}{(40-21)}$

35. a) $(20+11)!/11!$ (stack poles end on end with 11 identical) dividers
b) $\binom{(20-12)+12-1}{(20-12)}$ $\cdot 20!$

38. $\sum_{k=0}^{6}\binom{k+2-1}{k}\binom{(12-2k)+2-1}{(12-2k)}$ 41. $\sum_{k=0}^{7}\binom{k+3-1}{k}\binom{(20-k)+4-1}{(20-k)}$

45. a) $\sum_{k=0}^{10}\binom{(20-2k)+m-1}{(20-2k)}$ b) $\sum_{k=0}^{10}\binom{20}{k}\binom{20-k}{k}(m-2)^{20-2k}$

50. $25!/10!5!^3 + [(25!/12!6!7!)(2^7-2) - 25!/12!6!6!1!]\cdot 3 +$
$(25!/14!7!4!)2^4\cdot 3 + (25!/16!8!1!)\cdot 3!$;   first look at different
ways to write 25 as a sum of 4 successive integers, the first
is twice the size of next largst; the case of 1st integer = 12
& second largest integer = 6 & other 2 adding to 7 is the expansion
in brackets $(-25!/12!6!6!1!$ corrects for the fact that the
distribution 12,6,6,1 would be counted twice).

Section 5.5

1. a) $\binom{n-1}{k} + \binom{n-1}{k-1} = \dfrac{(n-1)!}{k!(n-1-k)!} + \dfrac{(n-1)!}{(k-1)!(n-k)!} =$

$\dfrac{(n-1)!(n-k)+(n-1)!k}{k!(n-k)!} = \dfrac{(n-1)!n}{k!(n-k)!} = \binom{n}{k}$

b) $\dfrac{n!}{m!(n-m)!} \cdot \dfrac{m!}{k!(n-k)!} = \dfrac{n!}{k!(n-k)!} \cdot \dfrac{(n-k)!}{(m-k)!(n-m)!}$

2 a) same reasoning as Example 2 but with right and left interchanged.
b) same reasoning as Example 3 but with m blocks followed by n blocks.

3. a) to pick r people from a row of n+r+1 people (right-hand side of (6)): pick all of the first r - i people, skip the (r-i+1)-st person, and pick i of remaining n+i people $(1 \cdot \binom{n+i}{i}$ ways).

b) to pick r + 1 people from a row of n+1 (right-hand side of (7)): pick r out of first i people and then pick (i+1)-st person (and skip remaining people), for i=r,r+1,...,n.

4. a) Induction step: $\sum_{k=0}^{n} \binom{n}{k} = \sum_{k=0}^{n}[\binom{n-1}{k} + \binom{n-1}{k-1}] = \sum_{k=0}^{n-1}\binom{n-1}{k} + \sum_{k=0}^{n-1}\binom{n-1}{k-1} =$

$2^{n-1} + 2^{n-1} = 2^n$

b) Induction step: $\sum_{k=0}^{r}\binom{n+k}{k} = \sum_{k=0}^{r-1}\binom{n+k}{k} + \binom{n+r}{r} = \binom{n+r}{r-1} + \binom{n+r}{r} =$

$\binom{n+r+1}{r}$.

5. Induction step: $(1+x)^n = (\sum_{k=0}^{n-1}\binom{n-1}{k}x^k(1+x) =$

$$\sum_{k=0}^{n-1}\binom{n-1}{k}x^k + \sum_{k=1}^{n}\binom{n-1}{k-1}x^k = \sum_{k=0}^{n}\binom{n}{k}x^k.$$

6. a) $\binom{-z}{k} = \frac{(-z)(-z-1)\cdots(-z-k+1)}{k!} = (-1)\binom{z+k-1}{k}$,

b) $\binom{z-1}{k} + \binom{z-1}{k-1} = \frac{(z-1)(z-2)\cdots(z-k+1)(z-k)}{k!} + \frac{(z-1)(z-2)\cdots(z-k+1)}{(k-1)!} =$

$\frac{(z-1)(z-2)\cdots(z-k+1)}{k!}[(z-k)+k] = \binom{z}{k}$.

8. a) same proof as 4b   9. $\binom{2n}{n} + \binom{2n}{n-1} = \binom{2n+1}{n} = 1/2[\binom{2n+1}{n} + \binom{2n+1}{n+1}] =$

10. n = 4 $=1/2\binom{2n+2}{n+1}$.

11. a) To pick 2 from n men and n women, either pick 2 of same sex or 1 of each sex.

14. a) $72\binom{n+2}{4}$ 15.a) $\sum_{k=0}^{n}(k+1)\binom{n}{k} = \sum\binom{n}{k} + \sum k\binom{n}{k} = 2^n + n2^{n-1} = (n+2)2^{n-1}$.

16. a) 0 ,   b) $n(n-1)2^{n-2}$,   c) 3^n

29. c) $1 - 2x - 2x^2 - 4x^3$.

30. a) Pick largest m_k with $\binom{m_k}{k} \leq n$, and recursively pick largest m_j with $\binom{m_j}{j} \leq n - \sum_{i=j+1}^{k}\binom{m_i}{i}$. Problem is, perhaps for some j, $m_{j-1} \geq m_j$.

If so, then $n - \sum_{i=j+1}^{k}\binom{m_i}{i} \geq \binom{m_j-1}{j-1} + \binom{m_j}{j} \geq \binom{m_j}{j-1} + \binom{m_j}{j} = \binom{m_j+1}{j}$. But this contradicts the maximality of m_j.

b) Let j be largest subscript for which m_j is not unique. Let m_j equal m or m', m< m'. Thus $\sum_{i=1}^{j}\binom{m_i}{i} \leq \sum_{i=0}^{j-1}\binom{m-i}{j-i} \overset{(6)}{=} \binom{m+1}{j}-1 < \binom{m'}{j}$, when $m_j = m$ and $m_{j-i} \leq m-i$, $i < j$.

Section 5.6

1. 1234, 1243, 1324, 1342, etc.   2. aceh, ache, aech, aehc, etc.
3. 1, 12, 123, 1234, 124, 13, 134, 14, 2, 23, 234, 24, 3, 34, 4, $\emptyset$
4. a, af, afg, afgh, afh, ag, agh, ah, f, fg, fgh, fh, g, gh, h, $\emptyset$
5. 1234,1235,1236,1245,1246,1256,1345,1346,1356,1456 ,2345,2346,
   2356,2456,3456
6. adh, adj, ahj, dhj
7. a) 1234··· 9(10),..... ,(10)987··· 21 , b) 1, ,...., n
   c) abcd, ...   , wxyz

Chapter VI

Section 6.1

1. a) 8 products: $xxxx, x^21xx, x^2x1x, x^2xx1, 1x^2xx, xx^21x, xx^2x1, x^2x^211$.
   b) 15 products: $11x^4, 1xx^3, 1x^2x^2, 1x3x, 1x^41, x^21x2, x^31x, x^411, xx^31,$
      $x^2x21, x^3x1, x^2xx, xx^2x, xxx^2, x1x^3$
   c) 10 products: $x^4111, 1x^411, x^2x^211, x^21x^21, x^21xx, x^211x^2, 1x^2x^21,$
      $1x^2xx, 1x^21x^2, 11x^2x^2$
   d) 15 products: $x^411, x^3x1, x^31x, x^2x^21, x^2xx, x^21x^2, xx^31, xx^2x, xxx^2,$
      $x1x^3, 1x^41, 1x^3x, 1x^2x^2, 1xx^3, 11x^4$.
2. a) $(1+x+x^2+x^3+x^4)^5$, b) $(x+x^2+x^3)^3$, c) $(x^3+x^5+x^7)(x^2+x^4+..x^8)(x^2+..x^8)^2$
3. a) $(1+x+x^2+x^3)(1+x+..x^4)^2$, b) $(x+x^2..x^5)(x+x^2+x^3)(x+x^2...x^8)$,
   c) $(1+x+x^2+...)^4$, d) $(x+x^3+x^5 ...)^2(1+x+x^2+ ...)^5$.
4. a) $(1+x+x^2+x^3+x^4)^5$, b) $(x^3+x^4+...x^8)^4$
   c) $(x+x^2+x^3+ ...)^7$, d) $(1+x+x^2+x^3+x^4+x^5)(1+x+x^2+ ...)^2$.
5. $(1+x+x^2+ ...)^4(1+x)^2$.coefficient of x^5.   6. $(1+x+x^2+ ...)^5$.
7. $(x^3+x^4+x^5+ ...)^4$, coefficient of x^{18}
8. a) $(1+x+ ...)^4$, coef. of x^{27}, b) $(x+x^2+...)^4$, coef. of 27,
   c) $(1+x+...+x^{13})^4$, coef. of x^{27}
9. $(1+x+...)^n$ 10. $(1+x)^u(1+x+x^2)^v(1+x+x^2+x^3)^w$.
11. $(1+x^2+x^4+..)(x+x^3+x^5+...)(1+x+x^2+...)^{n-2}$
12. $(x^{r_1}+x^{r_1+2}+...x^{s_1})(x^{r_2}+x^{r_2+2}+x^{s_2})(1+x+x^2+x^3)^{r-2}$.
13. $(x+x^2+... x^6)^n$ 14.a) $(x+x^3+x^5)^3(x^2+x^4+x^6)^3$, b) $\prod^6[(x+x^2..x^6)-x^i]$.
15. $(x^{-3}+x^{-2}+...x^3)^4 = x^{-12}(1+x+...x^6)^4$.   16.$(1+x^{i=1}x^9)^6$.
17. $(1+x^5+x^{10}+...)^8$ 18. a) $(x^2+x^3+..x^7)^5$, x^{20}, b) $x^{10}(1+x^2...x^5)^5$.
19. $(1+x+x^2+...)(x^2+x^3+...)^4(1+x+x^2+...)$, coef. of x^{15} or x^{n-5}; the
    n-5 non-selected integers are distributed among 6 "boxes" before,
    between, and after the chosen integers.
20. r is a predetermined value instead of an index.
27. $(xy+yz+xz)^n$.

Section 6.2

1. $\binom{n+8-1}{8}$ 2. $\binom{r-33}{r-40}$ 3. $\binom{m}{9}+\binom{m}{7} + \binom{m}{5}$
4. $\binom{9+6-1}{9}-\binom{4+6-1}{4}$ 5. $\binom{10+4-1}{10}-4\binom{4+4-1}{4}$ 6. $\binom{19+3-1}{19}-\binom{4+3-1}{4}-\binom{3+3-1}{3}$
7. a) $\binom{10+10-1}{10}$, b) $\binom{10+4-1}{10}-3\binom{9+4-1}{9}$, c) $\sum_{k=0}^{5}\binom{5}{k}\binom{(12-k)+5-1}{(12-k)}$, e) $\binom{12}{m}b^{12}$
8. a) $(1-x^{28})/(1-x^4)$, b) $(x^{20}-x^{200})/(1-x^{20})$ 9. 0.   10. $\binom{6+3-1}{6}$
11. a) 0, b) 1, c) $\binom{12+8-1}{12}$, d) $\binom{12+5-1}{12}4^{12}$, e) $\binom{4+4-1}{4}$
12. $10!/2!3!5!$.   13. a) $\binom{4+3-1}{4}$, b) $\binom{10+3-1}{10} - \binom{7+3-1}{7}$
14. a) $\binom{r-1}{r-8}$, b) $\binom{r/2+7}{r/2}$ if r even (=0, o/w)
15. $\binom{12+5-1}{12}-\binom{5}{1}\binom{7+5-1}{7}+\binom{5}{2}\binom{2+5-1}{2}$.   16. $\binom{10+5-1}{10}-3\binom{6+5-1}{6}+3\binom{2+5-1}{2}$
17.a)$\binom{15+n-1}{15}+\binom{10+n-1}{10}$, b) $\binom{15}{n}+\binom{10}{n}$,

17. c) $\sum_{k=0}^{5} (-1)^k \binom{(15-3k)+n-1}{(15-3k)} \binom{n}{k} + \sum_{k=0}^{3} (-1)^k \binom{(10-3k)+n-1}{(10-3k)} \binom{n}{k}$

18. $\binom{15+10-1}{15} - \binom{10}{1}\binom{9+10-1}{9} + \binom{10}{2}\binom{3+10-1}{3}$

24. $(1/\binom{25}{8})$[coef. of x^{17} in $(1+x+..+x^5)^9$], like Exer.19 in sect.3.1;
    $(1/\binom{25}{8})[\binom{17+9-1}{17} - \binom{9}{1}\binom{11+9-1}{11} + \binom{9}{2}\binom{5+9-1}{5}]$

19. $\binom{5+7-1}{5}$
    22. $1/2[\binom{15+6-1}{15} - \binom{6}{1}\binom{9+6-1}{9} + \binom{6}{2}\binom{3+6-1}{3}]$

27. Calculate coef. of x^r on each side of $(1+x)^m(1+x)^n = (1+x)^{m+n}$

38. c) $(\frac{1}{2})^5(1-\frac{1}{2})^{-5}$.   39.a) $p_X'(1)=[\sum_{r=0}^{\infty} rp_r t^{r-1}]_{t=1} = \sum rp_r$,   b)c)10

Section 6.3

1. a) 5 partitions: 4, 3+1, 2+1+1, 2+2, 1+1+1+1
   b)11 partitions: 6,5+1,4+2,4+1+1,3+3,3+2+1,3+1+1+1,2+2+2,
        2+2+1+1,2+1+1+1+1,1+1+1+1+1+1
2. a) $[(1-x^2)(1-x^4)(1-x^6)...]^{-1}$,   b) $(1+x)(1+x^3)(1+x^5)...$
3. $(1+x+x^2+x^3)(1+x^2+x^4+x^6)(1+x^3+x^6+x^9)...$
4. a) $[(1-x^2)(1-x^3)(1-x^7)]^{-1}$,   b) $\frac{x^{16}}{(1-x^2)}(x^6+x^9+x^{12}..x^{24})(1+x^7+x^{14})$
5. $[(1-x)(1-x^5)(1-x^{10})(1-x^{25})]^{-1}$
6. $[(1-x^2)(1-x^5)(1-x^{10})]^{-1}$
7. a) $[(1-x)(1-x^2)(1-x^3)]^{-1}$,   b) $[(1-x)(1-x^2)(1-x^3)...(1-x^n)]^{-1}$
8. a) 10 such partitions of 10 in each case.
   b) $(1+x)(1+x^2)... = (\frac{1-x^2}{1-x})(\frac{1-x^4}{1-x^2})(\frac{1-x^6}{1-x^3})(\frac{1-x^8}{1-x^4})... = [(1-x)(1-x^3)...]^{-1}$
14. Mimic argument in Example 4 of this section.
15. Use Ferrars graph: m rows of n dots, transpose is n rows of m dots.
19. b) Ferrars graph for r+k into k parts: write k dots in first
    column, then distribute remaining r dots to form a proper Ferrars
    graph; for r into parts of size $\le k$, delete first column in previous
    Ferrars graph and transpose.

Section   6.4

1. $(1+x^2/2! + .. x^5/5!)^5$.   2. $(x^2/2!+x^3/3!+x^4/4!)^6$
3. $(1+x)^5 e^{21x}$.   4.a) $(e^x-1)^n$,   b) $s_{n,r}=(1/n!)\sum_{k=0}^{n}(-1)^{n-k}\binom{n}{k}k^r$
5. coef. of $x^{13}/13!$ in $(1+x+x^2/2!+x^3/3!+x^4/4!)^{13}$
6. $4^8 - 3^8 - 8\cdot3^7$.   7. a) $1/2(3^r+1)$, b) $1/4(3^r+2+(-1)^r)$, 8.$\frac{1}{2}(4^r-3^r-2^r)$
9. a)$(1+x)^4 e^{22}$: $\sum_{k=0}^{4} \binom{4}{k}P(10,k)22^{10-k}$,   b)$(e^x-1)^4 e^{22}$:$\sum_{k=0}^{4}(-1)^k\binom{4}{k}(22-k)^{10}$
10. a) $(e^x-x^2/2!)^3$; $3^r-3\binom{r}{2}2^{r-2}+3\binom{r}{2}\binom{r-2}{r}$.   11. $\frac{1}{2}4^r$
15. a) e^x/x.   20. $a_r = \sum_{k=0}^{r}\binom{k+n-1}{k}n!/(n-k)!$

Section 6.5

1. a) $x/(1-x)^2$, b) $13/(1-x)$, c) $3x(1+x)/(1-x)^3$,
   d) $3x/(1-x)^2 + 7/(1-x)$, e) $24x^4/(1-x)^5$
2. a) $\binom{n+1}{2}$, b) $13n+13$, c) $3\binom{n+2}{3}+3\binom{n+1}{2}= \frac{1}{2}n(n+1)(2n+1)$
   d) $3\binom{n+1}{2}+7(n+1)$, e) $24\binom{n+1}{5}$
3. $(1/x)\frac{d}{dx}(x^2(x\frac{d}{dx}(1/(1-x))) = \frac{x(3-x)}{(1-x)^3}$.   5. a) $x^2(1+x)/(1-x)^3$

Chapter VII

Section 7.1

1. $a_n = 5a_{n-1}$, $a_1 = 5$.

2. a) $a_n = a_{n-1} + a_{n-2} + a_{n-3}$,   b) $a_5 = 13$

3. $a_n = a_{n-1} + 2a_{n-2}$.

4. a) $a_n = a_{n-3} + a_{n-6} + a_{n-10}$,   b) $a_{33} = 148$

5. $a_n = 3a_{n-1} + 2a_{n-5} + a_{n-10} + a_{n-25}$.

6. a) $a_n = a_{n-1} + a_{n-2}$,   b) $a_n = 2a_{n-1} + 2a_{n-2}$

7. $a_n = a_{n-1} + a_{n-2}$ (also 9,10).

8. identity $\binom{n}{k} = \binom{n-1}{k} + \binom{n-1}{k-1}$ and induct

12. $a_n = a_{n-1} + 2(n-1)$

13. a) $a_n = a_{n-1} + 1$, $n \le k$ or $a_n = a_{n-1} + n$, $n > k$

14. $a_n = 1.06(a_{n-1} + 50)$

16. $a_n = a_{n-1} + n - 1$.

17. $a_n = 4a_{n-1} + 2^{n-1}$

18. $a_n = 2a_{n-1} + 1$.

20. $a_n = (2n-1)a_{n-1}$.

22. $a_{n,k} = a_{n-1,k} + a_{n,k-1}$

26. Let $a_{n,k,i}$ = no. of such sequences starting with digit i, i=1,2:

29. a) $a_n = 2a_{n-1} + 4^{n-1}$ $a_{n,k,0} = a_{n-1,k,1}$, $a_{n,k,1} = a_{n-1,k,0} + a_{n-1,k-1,1}$

45. $a_n = a_{n-3}$, n even; $a_n = a_{n-3} + \lfloor \frac{n+1}{4} \rfloor$ ← n odd (round down to nearest int.)

Section 7.2

1. a) $An - 5/2$,   b) $a_n = 2n$.

2. $a_n = 2a_{n/2} + 1$, $a_n = n-1$

3. a) $a_n = a_{n/10} + 1$, $a_n = \lceil \log_{10} n \rceil$,   b) $a_n = 10a_{n/10} + 1$

4. $a_n = (1/9)(n-1)$.

5. $a_n = 2a_{n/2} + n-1$, $a_n = 0(n\log_2 n)$

Section 7.4

1. $a_n = (1.08)^n 500$.

2. $a_n = 2a_{n-1}$, $a_n = 3 \cdot 2^{n-1}$

3. a) $a_n = (\frac{2}{5})4^n + (3/5)(-1)^n$

5. $a_n = 2a_{n-1} + a_{n-2}$, $a_n = \frac{1}{4}(2+\sqrt{2})(1+\sqrt{2})^n - \frac{1}{4}(2-\sqrt{2})(1-\sqrt{2})^n$

6. $a_n = 2a_{n-1} + 2a_{n-2}$, $a_n = (1/6)(3+2\sqrt{3})(1+\sqrt{3})^n + (1/6)(3-2\sqrt{3})(1-\sqrt{3})^n$

7. $P_n - P_{n-1} = 2(P_{n-1} - P_{n-2})$, $P_n = 3 \cdot 2^n - 2$.   12. $C_1 = 9$, $C_2 = -18$

Section 7.5

1. a) $a_n = 1 + 3\binom{n}{2}$,   b) $g(x) = 3 + 2x/(1-x)^3$: $a_n = 3 + 2\binom{n+2}{3}$

2. $a_n = a_{n-1} + 2$, $a_n = 2n$.   5. $a_n = a_{n-1} + \sum_{k=1}^{n-3}(k(n-2-k)+1)$, $a_n = \binom{n}{4} + \binom{n}{2} - n + 1$

8. a) $a_n = 1.08 a_{n-1} + 10n$,   b) $a_n = (1000 + 10.8/.0064)(1.08)^n - 10n/.08 - .08/.0064$

9. a) $a_n = (5/3)2^n + (1/3)(-1)^n$

10. $a_n = 2a_{n-1} - a_{n-2} + 10 \cdot 2^n$, $a_n = 960n - 20 + 40 \cdot 2^n$

Section 7.6

1. a) $1/(1-x) + 2/(1-x)^2$,   b) $1/(1-x) + 2x^2/(1-x)^3$

   c) $(1-2x)/(1-3x+2x^2)$,   d) $1/(1-2x)^2$

7. a) $a_n = \sum_{k=3}^{n-1} a_{n+2-k} a_k$, $g(x) = (x/2)(1-\sqrt{1-4x})$, $a_n = \frac{1}{n-1}\binom{2n-4}{n-2}$

8. $a_n = 3a_{n-1}$, $n > 3$, $a_0 = a_1 = a_2 = 0$, $a_3 = 1$, $g(x) = x^3/(1-3x)$

9. $a_n = \sum_{k=1}^{n-2} a_k a_{n-k-1}$, $a_1 = 1, a_2 = 0$, $a_n = 0$, n even; $a_n = \frac{2}{n+1}\binom{n-1}{(n-1)/2}$ odd
         (hint: let $b_k = a_{2k+1}$; for b's, $h(x) = \frac{1}{2}(1 - \sqrt{1-4x})$.)

10. a) $a_n = \sum_{k=0}^{n-1}\binom{n-1}{k} a_{n-k-1}$,   b) $g(x) = e^{e^x - 1}$

11. $F_n(x) = F_{n-1}(x) + xF_n(x)$, $F_n(x) = (1-x)^{-n}$

Chapter VIII

Section 8.1

1. $2 \cdot 5 \cdot 26^4 - 5^2 \cdot 26^3$.

2. $10^9 - 10!$

3. $3^n - 3 \cdot 2^{n-1}$

4. $2^8 - \binom{8}{0} - \binom{8}{1}$.

5. $\binom{52}{5} - \binom{13}{5} 4^5$

6. $(n! - 1)/n!$

7. $60 + 30 - 20 = 70$

8. a) $200 - 100 - 70 + 30 = 60$, b) $(200 - 100) - 60 = 40$

9. $600 - 200 - 200 - 100 = 100$

10. a) $2 \cdot 24! - 22!$, b) $26! - 24! - 23! + 22!$

11. $10! - 2 \cdot 9! + 8!$

12. $4 \cdot 5^3 - 5 \cdot 4(4!/2)$ (all codes 1 red)
    (1 red, 2 of another color)

13. a) 35, b) 35.

14. a) yes: let A_1 = tennis, A_2 = bridge, A_3 = Master-
    mind, then $50\% \doteq N(A_3) < N(A_3 \cap (A_1 \cup A_2)) = N(A_1 A_3) + N(A_3 A_2) - N(A_1 A_2 A_3)$
    $= 45\% + 45\% - 20\% = 70\%$, b) 45%

15. $3^{20} - 3 \cdot 2^{20} + 3$

17. $6!/2!2!2! - 3 \cdot 5!/2!2! + 3 \cdot 4!/2! - 3!$

21. $3 \cdot 6!/2! - (2 \cdot 6!/3! + 6!/2!2!) + 6!/4!$

22. $[2\binom{11}{4}7!/2! + \binom{11}{3}8!/2!2!] - [\binom{11}{6}5! + \binom{11}{5}6!/2! + \binom{11}{4}\binom{7}{3}4!] + \binom{11}{7}4!$
    (pick places for 2 Ts and 2 Hs) (arrange other letters in other
    places)

23. $15!/3!^5 - 3 \cdot 5!10!/2!^5 + \binom{3}{2}5!^3 - 5!$.

26. n = 100

28. Ans. = 2: $N(\overline{GHL}) = N - \overline{N(G) + N(H) + N(L)} + \overline{N(GH) + N(GL) + N(HL)} - \overline{N(GHL)} \Rightarrow$
    $N(GH) + N(GL) + N(HL) = 20$, i.e. everyone takes 2 languages.
    So $N(GH) = 22 - N(L) = 2$

29. 45

Section 8.2

1. $10^n - 3 \cdot 9^n + 3 \cdot 8^n - 7^n$.

2. $\sum_{k=0}^{6} (-1)^k \binom{6}{k} (6-k)^{10}$.

3. $13\binom{48}{6} - \binom{13}{2}\binom{44}{2}$

4. $420[1 - (1/2 + 1/3 + 1/5 + 1/7) + (1/2 \cdot 3 + 1/2 \cdot 5 + 1/2 \cdot 7 + 1/3 \cdot 5 + 1/3 \cdot 7 + 1/5 \cdot 7)$
   $- (1/2 \cdot 3 \cdot 5 + 1/2 \cdot 3 \cdot 7 + 1/2 \cdot 5 \cdot 7 + 1/3 \cdot 5 \cdot 7) + 1/2 \cdot 3 \cdot 5 \cdot 7]$

5. a) $(1/\binom{52}{13})[\binom{52}{13} - \binom{4}{1}\binom{39}{13} + \binom{4}{2}\binom{26}{13} - \binom{4}{3}]$, b) $1 - $ (answer in part a),
   c) $(1/\binom{52}{13})[\binom{52}{13} - \binom{4}{1}\binom{48}{13} + \binom{4}{2}\binom{44}{13} - \binom{4}{3}\binom{40}{13} + \binom{4}{4}\binom{36}{13}]$

6. $10!/2^5 - \binom{5}{1}5 \cdot 8!/2^4 + \binom{5}{2}5 \cdot 4 \cdot 6!/2^3 - \binom{5}{3}5 \cdot 4 \cdot 3 \cdot 4!/2^2 + \binom{5}{4}5! - \binom{5}{5}5!$

7. $9!/3!^3 - \binom{3}{1}7!/3!^2 + \binom{3}{2}5!/3! - 3!$

8. $\sum_{k=0}^{n} (-1)^k \binom{n}{k} (2n-k)!/2^{n-k}$

9. $26! - (3 \cdot 23! + 24!) + (2 \cdot 20! + 2 \cdot 21!)$; R̲U̲N & F̲R̲O̲M, F̲R̲O̲M & J̲O̲E̲

10. a) $\binom{30+4-1}{30} - \binom{4}{1}\binom{19+4-1}{19} + \binom{4}{2}\binom{8+4-1}{8}$

11. $\sum_{k=0}^{3} (-1)^k\binom{3}{k}\binom{(25-7k)+6-1}{(25-7k)}$
12. a) $\sum_{k=0}^{4} (-1)^k\binom{4}{k}\binom{(12-2k)+(4-k)-1}{(12-2k)}$

13. $\sum_{k=0}^{n} (-1)^k\binom{n}{k}\binom{(r-n-5k)+n-1}{(r-n-5k)}$
16. $\approx 26!e^{-1}$.
17. $\approx (10!)^2e^{-1}$

18. $\sum_{k=0}^{n/2} (-1)^k\binom{n/2}{k}(n-k)!$
20. $(10!/2!^5)\sum_{k=0}^{5} (-1)^k\binom{5}{k}(10-2k)!/2^{5-k}$

21. a) $\sum_{k=0}^{n-1} (-1)^k\binom{n-1}{k}(n-k)!$
b) Use identity (1) of section 2.6.

25. The subproblems of counting $N(A_i)$, $N(A_iA_j)$, and $N(A_iA_jA_k)$ are each themselves inclusion-exclusion problems:

$9!/3!^3 - 3[8!/3!^2-7!/3!] + 3[7!/3!-2\cdot6!/3!+5!/3!]-[6!-3\cdot5!+3\cdot4!-3!]$

27. $P(\binom{7}{3},7) - \binom{7}{1}P(\binom{6}{3},7) + \binom{7}{2}P(\binom{5}{3},7)$

30. $5^5 - \binom{8}{1}5^4 + \binom{8}{2}5^3 - (4\cdot5^3+(\binom{8}{3}-4)5^2) + (21\cdot5^2+(\binom{8}{4}-21)5)$
$- (4\cdot5^2+(\binom{8}{5}-4)5) + 21\cdot5 = 340$

Section 8.3

1. darkened squares are on the main diagonal

2. a) $1 + 8x + 21x^2 + 20x^3 + 6x^4$, c) $1 + 7x + 11x^2 + 4x^3$,
   d) $1 + 8x + 14x^2 + 4x^3$

3. $7! - 9\cdot6! +30\cdot5! - 46\cdot4! +32\cdot3! - 8\cdot2! = 1232$

4. $(1/29^6)(6!-7\cdot5!+18\cdot4!-21\cdot3!+11\cdot2!-2\cdot1!)6!$ (ways to order the 6 pairs of values)

5. a) $(4!-5\cdot3!+8\cdot2!-4\cdot1!)/(5!-8\cdot4!+22\cdot3!-25\cdot2!+12\cdot1! - 2\cdot0!)= 6/20$
6. $5! - 7\cdot4! + 16\cdot3! - 13\cdot2! +2\cdot1! = 24$ b) $12/20$

Chapter IX

Section 9.1

1. all No, except b .   2. all Yes, except a .

3. a) 6 symmetries: $\binom{abc}{abc}$, $\binom{abc}{bca}$, $\binom{abc}{cab}$, $\binom{abc}{acb}$, $\binom{abc}{cba}$, $\binom{abc}{bac}$,

   b) 4 symmetries,   c) one symmetry- the identity

4. a) (abcd), b) (ac)(bd), c) (ad)(bc), d) (a)(b)(c)(d),
   e) (ab)(cd), f) (ad)(bc), g) (1346)(257)

5. a) all C_i fixed,   b) $\binom{C_1 C_2 C_3 C_4 C_5 C_6 C_7 C_8 C_9 C_{10} C_{11} C_{12} C_{13} C_{14} C_{15} C_{16}}{C_1 C_3 C_4 C_5 C_2 C_7 C_8 C_9 C_6 C_{11} C_{10} C_{13} C_{14} C_{15} C_{12} C_{16}}$

6. a) 2^3 colorings

   b) (i) $\binom{T_1 T_2 T_3 T_4 T_5 T_6 T_7 T_8}{T_1 T_3 T_4 T_2 T_6 T_7 T_5 T_8}$,   (ii) $\binom{T_1 T_2 T_3 T_4 T_5 T_6 T_7 T_8}{T_1 T_2 T_4 T_3 T_7 T_6 T_5 T_8}$

7. see Exer. 8.

8. a) Rotations are mutually distinct (each take a given corner a
   to a different corner), same for flips, and rotation≠ flip since
   only flips invert cyclic order of corners.
   b) A given corner a can go to any of n corners, denoted a*, and
   edge (a,b) can be on the clockwise or counter-clockwise side of a*.
   A corner and an edge totally determine a symmetry⇒2n symmetries.

10. a) π_1 , b) π_8, c) π_2, d) π_2

12. Any flip composed with any rotation.

Section 9.2

1. a) 10, b) 24, c) 70 .   2. $1/3(3^4 + 2\cdot 3^2) = 33$

3. $1/3(3^{15} + 2\cdot 3^5)$.   4. $1/4(2^{64}+2\cdot 2^{16}+2^{32})$.   5. $1/5(3^5+4\cdot 3)=51$

6. a) $1/2(7^2+7)= 56$, b) $\binom{2+7-1}{2}= 28$.

7. $1/2(5^n+ 5^{n/2})$, n even; $1/2(5^n+ 3\cdot 5^{(n-1)/2})$, n odd

8. $1/4(4^5+ 2\cdot 4^2+ 4^3)= 280$.   10. $1/2(3\cdot 2^{n-1}+3\cdot 2^{(n-1)/2})$, n odd

                                  $1/2(3\cdot 2^{n-1})$, n even ($\Psi(180°)=0$)

11. $\Upsilon(\pi_1)= 18, \Upsilon(\pi_3)= 6, \Upsilon(\pi_7)= \Upsilon(\pi_8)= 12$, and other Υ's = 0;
    answer $1/8(18+6+12+12)= 6$

Section 9.3

1. $1/8(4^4 + 2 \cdot 4 + 3 \cdot 4^2 + 2 \cdot 4^4 \cdot 4) = 55$

2. a) $1/5(4^5 + 4 \cdot 4) = 208$,   b) $1/10(4^5 + 5 \cdot 4^3 + 4 \cdot 4) = 136$

3. a) $1/6(3^6 + 3^3 + 2 \cdot 3^2 + 2 \cdot 3) = 130$,   b) $1/12(3^6 + 3 \cdot 3^4 + 4 \cdot 3^3 + 2 \cdot 3^2 + 2 \cdot 3) = 92$,
      c) the cyclic color sequences of two such hexagons are
         R-W-B-R-W-B- and R-B-W-R-B-W-

4. a) $1/7(3^7 + 6 \cdot 3) = 315$,   b) $1/9(3^9 + 2 \cdot 3^3 + 6 \cdot 3) = 2195$,
   c) $1/10(3^{10} + 3^5 + 4 \cdot 3^2 + 4 \cdot 3) = 5934$,   d) $1/11(3^{11} + 10 \cdot 3) = 16{,}107$

5. a) $1/6(m^4 + 3m^3 + 2m^2)$,   b) $1/8(m^9 + 4m^6 + m^5 + 2m^3)$,   c) $1/2(m^5 + m^3)$

6. a) $1/8(x_1^4 + 2x_4 + 3x_2^2 + 2x_1^2 x_2)$, b) 21, c)$P_G = \frac{1}{8}(x_1^8 + x_2^4 + 2x_4^2 + 4x_1^2 x_2^3)$: 954,
   d) the relative position of colored edges with respect to colored
      corners creates additional nonequivalent colorings $\square \ne \square$

7. a) $1/6(m^6 + 3m^4 + 2m^2)$, b) $1/8(m^{12} + 2m^7 + 4m^6 + 2m^3)$,   c) $1/2(m^6 + m^4)$

8. a) Answer is $\psi(0°)/n$ for n equals 3 and 7 ; find $\psi(0°)$ by
      inspection or inclusion-exclusion formula: (i) 2, (ii) , (iii) 258

9. $1/4(7^4 + 3 \cdot 7^2) = 637$.   11. a) $1/2(2^8 + 2^4) = 136$

Section 9.4

1. $b^5 + b^4 w + 2b^3 w^2 + 2b^2 w^3 + bw^4 + w^5$

2. a) $b^6 + b^5 w + 3b^4 w^2 + 3w^3 + 3b^2 w^4 + bw^5 + w^6$,
   b) $b^9 + b^8 w + 4b^7 w^2 + 10b^6 w^3 + 14b^5 w^4 + 14b^4 w^5 + 10b^3 w^6 + 4b^2 w^7 + bw^8 + w^9$.
   c) $b^{10} + b^9 w + 5b^8 w^2 + 12b^7 w^3 + 22b^6 w^4 + 26b^5 w^5 + \cdots$ (by symmetry),
   d) $b^{11} + w^{11} + \sum_{k=1}^{10} \frac{1}{11}\binom{11}{k} b^{11-k} w^k$

3. $b^4 + w^4 + r^4 + b^3 w + b^3 r + w^3 b + w^3 r + r^3 b + r^3 w + 2b^2 w^2 + 2b^2 r^2 + 2w^2 r^2 + 2b^2 wr + 2w^2 br + 2r^2 bw$

4. $1/4[(b+w)^{16} + 2(b^4 + w^4)^4 + (b^2 + w^2)^8]$

5. a) $1/6[(b^4 + w^4) + 3(b^2 + w^2)(b+w)^2 + 2(b^3 + w^3)(b+w)]$,
   b) $1/4[(b+w)^6 + (b^2 + w^2)^2(b+w)^2 + 2(b^2 + w^2)^3]$

6. a) $b^5 + 2b^4 w + 3b^3 w^2 + 3b^2 w^3 + 2bw^4 + b^5$,
   b) $1/8[(b+w)^8 + 2(b^4 + w^4)^2 + 3(b^2 + w^2)^4 + 2(b^2 + w^2)^2(b+w)^4]$
   c) same as part a

7. a) $1/6[(b+w)^6 + 3(b+w)^2(b^2+w^2)^2 + 2(b^3+w^3)^2]$,

   b) $1/4[(b+w)^7 + 2(b+w)(b^2+w^2)^3 + (b+w)^3(b^2+w^2)^2]$

8. $1/24[(b+w)^{12} + 6(b^4+w^4)^3 + 3(b^2+w^2)^6 + 6(b+w)^2(b^2+w^2)^5 + 8(b^3+w^3)^4]$

9. a) same as corners (Example 3),

   b) $1/24[(b+w)^6 + 6(b+w)^2(b^4+w^4) + 3(b+w)^2(b^2+b^2)^2 + 6(b^2+w^2)^3 +$
   $$8(b^3+w^3)^2]$$

10. a) $b^2w^2r + b^2wr^2 + bw^2r^2$,   b) none possible,

   c) $b^4w^4 + b^4r^4 + w^4r^4 + b^4w^2r^2 + b^2w^4r^2 + b^2w^2r^4 + b^4w^3r + b^4wr^3 + b^3w^4r$
   $bw^4r^3 + b^3wr^4 + bw^3r^4 + b^3w^3r^2 + b^3w^2r^3 + b^2w^3r^3$

Chapter X Combinatorial Modeling in Theoretical Computer Science

Section 10.1

1. a) s =>1s => 101s => 1011s=> 101101s=> 101101
   b) s=> 2s=> 22s=> 220t=> 2201t=> 22010t=> 220102t=> 220102
   c) s=> $2s_2$=> $20t_3$=> $202t_4$=> $2020t_5$=> 20202

   d) s=> t_3p_8=> $c_{3s}c_{3h}c_{3c}p_8$=> $c_{3s}c_{3h}c_{3c}c_{8h}c_{8d}$

   e) s=> 0s1=> 00s11=> 000s111=> 00001111

2. a) s=> tt=> 0t=> 001t=> 0010
   b) s=> tt=> 10tt=> 100t=> 1001
   c) s=> tt=> 0t=> 001t=> 00110t=> 001100

3. a) s=> tt=> : (i) 01tt=> 010t=> 01010t=> 010101; (ii) 0t=>
      010t=> 01010t=> 010101; (iii) 01tt=> 0101tt=> 01010t=> 010101
   b) same type of derivations as in a.

4. s -> 1t0, t -> 1, t -> 0, t -> ∅.

5. s ->0t1v, t ->0, t ->∅, v ->1, v ->∅ (assumes at least one 0 &
   1, otherwise s -> tv).

6. s ->s_1, s_i ->$0s_{i+1}$, s_i ->$1s_{i+1}$, i=1,2,3,4,5, s_6 -> ∅

7. a) s-> pppp, p-> R, p-> W, p-> Bu, p-> Bk, p-> Y, p-> G
   b) s-> Rp'p'p', s-> p'Rp'p', s-> p'p'Rp', s-> p'p'p'R, where
      p' has same productions as p in a) except p' doesn't yield R.
   c) s-> Rppp, s-> p'Rpp, s-> p'p'Rp, s-> p'p'p'R, p has the
      productions in a). p' has the productions in b).

8. a) s-> f_is_j, for all i≠j, f_i-> $c_{is}c_{ih}c_{id}c_{ic}$, s_j-> c_{jx}, all x.

9. a) s-> s_6, s-> As_5, s-> AAs_4, s-> $AAAs_3$,

   s_6-> $BBBt_3$, s_5-> BBt_3, s_5-> $BBBt_2$, s_4-> Bt_3, s_4-> BBt_2,

   s_4-> $BBBt_1$, s_3-> t_3, s_3-> Bt_2, s_3-> BBt_1, s_3-> BBB,

   t_3-> CCC, t_2-> CC, t_1-> C.

11. s-> 0r, s-> 1q, s-> 0t', s-> 1t', t-> 0r, t->1q, r->1t, q-> 0t,
    t-> 0t', t-> 1t', t'-> 0r', t'-> 1q', r'-> 1t', q'-> 0t',
    t'-> 0, t'-> 1. Note: t' is the second t in s-> tt.

13. $s \to t_1$, $t_i \to Tt_{i+1}$, $i=1,2$, $t_i \to Hu_{i+1}$, $i=1,2$, $t_3 \to Hu_4$,

$u_i \to Tu_{i+1}$, $i=2,3$, $u_i \to Hv_{i+1}$, $i=2,3$, $u_4 \to Hv_5$,

$v_i \to Tv_{i+1}$, $i=3,4$, $v_5 \to H$.

Section 10.2

1. a) $(t_0,-) \overset{1}{\to} (s_1,1) \overset{1}{\to} (s_1,1) \overset{0}{\to} (t_1,1) \overset{1}{\to} (s_2,2) \overset{1}{\to} (s_2,2) \overset{0}{\to} (t_2,2)$

 b) $(t_0,-) \overset{1}{\underset{1}{\to}} (s_1,1) \overset{0}{\to} (t_1,1) \overset{1}{\to} (s_2,2) \overset{0}{\to} (t_2,2) \overset{1}{\to} (s_3,3) \overset{0}{\to} (s_3,3)$
$\to (s_3,3)$

2. a) $(s_0,-) \overset{1}{\to} (s_0,Y) \overset{0}{\to} (s_1,Y) \overset{1}{\to} (s_0,Y) \overset{0}{\to} (s_1,Y) \overset{1}{\to} (s_0,Y)$

 b) $(s_0,-) \overset{1}{\underset{1}{\to}} (s_0,Y) \overset{0}{\to} (s_1,Y) \overset{1}{\to} (s_0,Y) \overset{0}{\to} (s_1,Y) \overset{0}{\to} (s_2,N) \overset{1}{\to} (s_2,N)$
$\to (s_2,N)$

3. a) $(t_0,-) \overset{5}{\to} (t_1,6) \overset{-}{\to} (t_1,-) \overset{-}{\to} (t_1,-) \overset{3}{\to} (t_1,3) \overset{-}{\to} (t_1,-) \overset{-}{\to} (t_1,-)$

 $\overset{1}{\to} (t_1,1) \overset{-}{\to} (t_1,-) \overset{-}{\to} (t_1,-)$

 b) $(t_0,-) \overset{6}{\to} (d_1,-) \overset{-}{\to} (d_1,-) \overset{-}{\to} (d_1,-) \overset{5}{\to} (d_2,-) \overset{-}{\to} (d_2,-) \overset{-}{\to} (d_2,-)$

 $\overset{3}{\to} (s_5,6) \overset{-}{\to} (s_4,5) \overset{-}{\to} (t_1,4)$

 d) the next state is undefined when the last digit is read in.

4. a)

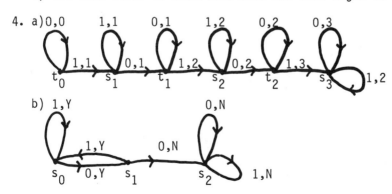

 b)

NOTE: In Exercises 5 α 7 The output and the next state names are the same.

5.

	-1	+1
0	-1	+1
+1	0	+2
+2	+1	+3
+3	+2	+3
-1	-2	0
-2	-3	-1
-3	-3	-2

6.

	0	1	2
s_0	$t_0,0$	$s_0,0$	$s_0,0$
t_0	$s_0,0$	$u_0,0$	$s_0,0$
u_0	$s_0,0$	$s_0,0$	$s_1,1$
s_1	$t_1,1$	$s_1,1$	$s_1,1$
t_1	$s_1,1$	$u_1,1$	$s_1,1$
u_1	$s_1,1$	$s_1,1$	$s_2,2$
s_2	$t_2,2$	$s_2,2$	$s_2,2$
t_2	$s_2,2$	$u_2,2$	$s_2,2$
u_2	$s_2,2$	$s_2,2$	$s_3,3$
s_3	$s_3,3$	$s_3,3$	$s_3,3$

7.

	0	1	2	3	4	5
0	0	1	2	3	4	5
1	1	2	3	4	5	0
2	2	3	4	5	0	1
3	3	4	5	0	1	2
4	4	5	0	1	2	3
5	5	0	1	2	3	4

8.

	0	1
s_0	$t_0,1$	$s_1,0$
s_1	$t_0,1$	$s_1,1$
t_0	$t_0,0$	$t_0,1$

9.

	0	1
s_0	$s_1,1$	$s_2,0$
s_1	$t_0,1$	$s_3,0$
s_2	$s_3,0$	$s_3,1$
s_3	$t_0,1$	$s_3,1$
t_0	$t_0,0$	$t_0,1$

11.

	0	1	2
s	$\{s,t\},0$	$r,0$	$r,0$
$\{s,t\}$	$\{s,t\},0$	$\{t,v\},0$	$r,0$
$\{t,v\}$	$\{s,t\},0$	$\{t,v\},0$	$r,*$
r	$r,0$	$r,0$	$r,0$

Section 10.3

1. $x_{i,1} \vee x_{i,2} \vee x_{i,3}$, $\bar{x}_{i,1} \vee \bar{x}_{i,2}$, $\bar{x}_{i,1} \vee \bar{x}_{i,3}$, $\bar{x}_{i,2} \vee \bar{x}_{i,3}$, $i = 1,2,3,4$

   $\bar{x}_{i,k} \vee \bar{x}_{j,k}$, $(i,j) = (1,2),(1,3),(2,3),(2,4),(2,5)$, $k = 1,2,3$

   Solution— $x_{1,1}$, $x_{2,2}$, $x_{3,3}$, $x_{4,1}$ True, other $x_{i,k}$ False.

2. $x_{i,1} \vee x_{i,2} \vee x_{i,3} \vee x_{i,4}$, $i = 1,2,3,4$, $x_{1,1}$

   $\bar{x}_{i,k} \vee \bar{x}_{j,k}$, for all pairs (i,j) and $k = 1,2,3,4$

   $\bar{x}_{i,k} \vee x_{j,k+1}$, for the following (i,j) pairs: $(1,2),(1,3),(2,1)$,
   $(2,3),(2,4),(3,1),(3,2),(3,4),(4,2),(4,3)$
   Solution: $x_{1,1}$, $x_{2,2}$, $x_{3,3}$, $x_{4,4}$ True, other $x_{i,k}$ False.

3. $(x_1 \wedge x_2 \wedge \bar{x}_3 \wedge \bar{x}_4) \vee (x_1 \wedge \bar{x}_2 \wedge x_3 \wedge \bar{x}_4) \vee (x_1 \wedge \bar{x}_2 \wedge \bar{x}_3 \wedge x_4) \vee (\bar{x}_1 \wedge x_2 \wedge x_3 \wedge \bar{x}_4)$

   $\vee (\bar{x}_1 \wedge x_2 \wedge \bar{x}_3 \wedge x_4) \vee (\bar{x}_1 \wedge \bar{x}_2 \wedge x_3 \wedge x_4)$

   Solution x_1, x_2 True, other x_i False.

4. In each graph t is connected to all $v_i, \bar{v}_i$; s to all a_i, b_i, c_i

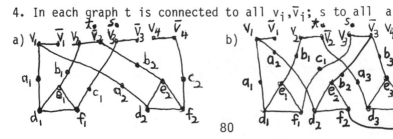

a)

b)

80

Coloring for a) and b): t color 2, s color 1, all v_i color 1, all $\bar{v}_i$ color 3. The proper colorings of other vertices are easily found

5. Obtain graphs for a) and b) by eliminating s and t and contracting the edges $(a_i,d_i),(b_i,e_i),(c_i,f_i)$.
   Edge covers: a) $v_1,e_1,f_1,v_2,e_2,f_2,v_3,v_4$

   b) $v_1,e_1,f_1,v_2,e_2,f_2,v_3,d_3,e_3,v_4$

6. Complement the graphs in 5 to get graphs for 6a) and 6b)
   a) want a $K_6:\bar{v}_1,\bar{v}_2,\bar{v}_3,\bar{v}_4,d_1,d_2)$ b) $K_7:f_3$ added to K_6 in a).

7. Similar to Example 1, (1) is now $x_{i,1} v x_{i,2}$, $\bar{x}_{i,1} v \bar{x}_{i,2}$.

8. Just consider the combined collection of disjunctions in Example 2 for each possible starting vertex.

9. In Example 2, let k run from 1 to n+1 and set $x_{1,n+1}$ True.

10. a) Define an $x_{i,j}$ for each edge (i,j). For each i, we want the disjunction $V(x_{i,j}^{\cdot,j} \wedge x_{i,k'})$,ranging over all $v_j,v_{j'}$ in $N(v_i)$

    and $\bar{x}_{i,j} v \bar{x}_{i,j'} v \bar{x}_{i,j''}$ for all $v_j,v_{j'},v_{j''}$ in $N(v_i)$.

    b) We also must forbid any subcircuit, i.e., for each subcircuit $(v_{i_1},v_{i_2},\ldots,v_{i_m},v_{i_1})$, we require

    $$\bar{x}_{i_1,i_2} v \bar{x}_{i_2,i_3} v \ldots v \bar{x}_{i_m,i_1}$$

Chapter XI

Section 11.1

1.

Look at possible factors for each of the three 4-edges.

2.

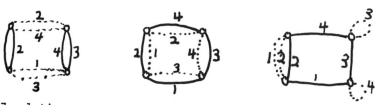

3. 7 solutions

4. Look at 1-edges: each circuit with a 1-edge between G,B and R,W must contain another non-1-edge between G,B and R,W. But only one such non-1-edge.

5. a) Yes, Hamiltonian circuit is a factor,
   b) No, ⋈

6. a) a-b-e-g-a, c-d-h-f-c,
   b) None,
   c) e-f-g-e, i-j-k-i, a-b-d-c-h-ℓ-a.

Section 11.2

1. a) a,c or b,d
   b) f,
   c) None exists; either h or s(h) = b in kernel. Either forces f in kernel. Now no way to complete kernel.

2. First player removes 1 on first move and subsequently always reduces pile size to a multiple of 6.

3. 3,4,9,11,12,16,17,21,25,26,27,31,32,36, over 40

4. If first player starts with 2, second wins at 4; if first player moves to 1 or 5, second to 6. Now if first player moves to 7 or 8, second wins at 9; if first player moves to 11, second wins at 16.

6. a) (red,blue) kernel points: (0,0),(1,2),(2,1),(3,5),(5,3),
      (4,7),(7,4),(6,10),
   b) (0,0),(1,2),(2,1),(3,5),(5,3),(0,6),(6,0),(6,6),(2,7),
      (7,2),(1,8),(5,9),(7,8).

7. a) g(c)=g(a)=0,g(b)=g(b)=1or interchange values,
   b) None exists (G-f is odd circuit).

82

7. c) g=0 for 3,4,9,11,12,16,17,21,25,26,27,31,32,36, over 40;
        g=1 for 0,1,5,6,10,14,15,18,19,23,24,29,30,34,35,39,40;
        g=2 for 2,7,8,13,20,28,33,37,38; g=3 for 20.

8. The vertices with $g(x)=0$ are the kernel, but we showed that
    this graph has no kernel.

11. a) Starting at x_1 successively pick as next vertex of the
        path a vertex whose level number is one smaller than
        level number of current vertex (always possible).

12. Proof easily by induction on level numbers.

14. Give kernel of G Grundy number 0, delete kernel. Give kernel
    of remaining graph Grundy number 1, and continue this process.
    Easy to verify by induction that this is proper Grundy func-
    tion.

15. Proceed as in construction in text. If at any stage there
    is a directed circuit C of unlabeled vertices x_1-x_2-x_3-...
    x_k-x_1, then find the x_i's in C with smallest potential
    Grundy number (smallest number not assigned to its successors
    that have already been labeled). If all x_i's have smallest
    potential Grundy number, pick any x_i and assign it that num-
    ber. Otherwise pick an x_j with smallest potential Grundy num-
    but x_{j+1} not, and give x_j this Grundy number. After
    breaking up a circuit of unlabeled vertices, continue as in
    text. If any difficulty later arises because of this method
    of breaking circuits, it must be that circuit was odd length
    (details omitted).

Section 11.3

1. a) 7, remove 3 from 4th pile,
    b) 0, second player can win,
    c) 4, remove 4 from 2nd, 3rd, or 4th pile,
    d) 0, second player can win.

2. a) 0, second player can win,
    b) 2, remove 2 from 3rd pile,
    c) 2, remove 2 from 1st pile,
    d) 1, remove 1 from 3rd pile.

3. a) 3, remove 3 from 3rd pile (or 2 from 4th pile),
    b) 2, remove 2 from 3rd or 4th pile ,
    c) 0, second player can win,
    d) 0, second player can win.

4. a) remove 3 from 2nd, 3rd, or 4th pile,
    b) remove 2 from 2nd, 3rd, or 4th pile,
    c) remove 1 from 2nd or 3rd pile or remove 5 from 4th pile.
5. a) 3, add nickel to 3rd pile.
9. a) remove a row of three, i.e.

APPENDICES

Appendix 1

1.a) {12,27}, b){2,3,6,7,9,12,15,17,18,21,22,24,27}, c){1,4,5,8, 10,11,13,14,16,19,20,23,25,26,28,29 }, d) all of A except 12,27.

2.a) A∩B∩C, b) $\overline{A∪B∪C}$, c) A∩B∩$\overline{C}$, d)not possible, cannot separate 2 and 4, e) not possible, cannot separate 6 and 10, f) $(\overline{A-(B∩C)})$

3. This information implies 2 calculators with memory are recharge-able and 2 non-rechargeable ones have no memory: So the 2 remaining calculators are rechargeable and have no memory.

4. Pick the 2 rechargeable memory ones and the 2 non-rechargeable no-memory ones (1 way to do this selection) or pick the 2 rechargeable no-memory ones and the 2 non-rechargeable memory ones (1 way).

5. a) Impossible information, we already know there are 7 men and 8 married women 7 + 8 ≠ 12.
   b) 4,   c) 5

6.

7.a)

b)

8.a) $\overline{A}∪B$ b)(A∩B)∪($\overline{A∪B}$)   17.b)(S∪H∪C)-(S∩H∩C)

Appendix 2

1. A=10; for I=1 to 20 do A←2xA; print A.

84

3. True for n=1. Induction step: $1^2 + 2^2 + \cdots + (n-1)^2 + n^2 =$

$$\frac{(n-1)(n)(2(n-1)+1)}{6} + \frac{6n^2}{6} = \frac{n}{6}[(2n^2-3n+1)+6n]$$

$$= \frac{n}{6}(2n^2+3n+1) = \frac{1}{6}n(n+1)(2n+1)$$

6. True for n=1. Induction step: $1^3 + 2^3 + \cdots (n-1)^3 + n^3 = \frac{1}{4}(n-1)^2 n^2 +$

$$\frac{4n^3}{4} = \frac{1}{4}n^2(n+1)^2$$

7. Induction step. $(1+2+\cdots n)^2 = (1+2+\cdots n-1)^2 + 2[\,1+2+\cdots n-1\,]n + n^2$

$$= 1^3 + 2^3 + \cdots (n-1)^3 + 2[\,1/2\,(n-1)n\,]n + n^2 =$$

$$1+2^2+\cdots(n-1)^3 + n^3$$

8. Induction step. $\frac{1}{1\cdot 2} + \cdots \frac{1}{(n-1)n} + \frac{1}{n(n+1)} = \frac{n-1}{n} + \frac{1}{n(n+1)} = \frac{(n^2-1)+1}{n(n+1)} = \frac{n}{n+1}$

11. Show $(1-a)(1+a+\cdots a^n) = 1-a^{n+1}$. Induction step

$$(1-a)[(1+a+a^2+a^{n-1})+a^n] = 1-a^n +(1-a)\,a^n$$

$$= 1 -a^n + a^n -a^{n+1} = 1-a^{n+1}$$

13. Induction step $a_n = 3a_{n-1} -2a_{n-2} = 3(2^n-1)-2(2^{n-1}-1)=2\cdot 2^n-1=2^{n+1}-1$.

16. Induction step $m\cdot n = [(m-1)+1]\cdot n = (m-1)n+n = n(m-1)+n = n(m-1)+n\cdot 1 = n[(m-1)+1]=n\cdot m$

18. Induction step. $x_1, \ldots, x_n$ alone have 2^{n-1} subsets. There are in addition 2^{n-1} subsets containing x_n plus some subset of $x_1, \ldots, x_{n-1}$. In total, $2^{n-1} + 2^{n-1} = 2^n$ subsets.

25. Induction step is false because when n=2, one cannot assume $x_{n+1} = x_n$.

Appendix 3

1. 4/8

3. a) 6/36 b) 18/36 c) 3/36
2. $P(E_0) = P(E_3) = 1/8$, $P(E_1) = P(E_2) = 3/8$

5. a) 1/6 b) 3/6
4. $(3\cdot 5 + 1)/6^3 = 16/216$

6. a) 1/10 b) 6/10

7. a) $3^2/9^2 = 1/9$ b) $(6\cdot 3+3\cdot 6)/9^2 = 4/9$ 8. $(1+3)/\binom{5}{2} = 4/10$

9. 2/6 10. $1-5^5/6^5$

11. $100-50-40+20 = 30$; 30/100

Appendix 4

1. $n+1$

2. If everyone has at least 1 friend (but no more than 19 friends), then of 20 numbers between 1 and 19, 2 numbers must be equal. If exactly 1 person has no friend, repeat the previous argument with the remaining 19 people. If 2 or more people have no friends, result is immediate.

3. No. Pigeonhole Principle does not apply: if n players, i-th player could win i-1 matches.

4. 52

5. If each person has 5 or more acquaintances, there are at least $\frac{1}{2}(5 \cdot 20) = 50$ pairs.

6. Smallest n with $\binom{n}{3} \geq 12$; $n = 6$.

7. If difference is not a multiple of 10, then each integer is different mod 10. If sum is not a multiple of 10, then at most one integer from each of the residue class $\{0\},\{1,9\},\{2,8\},\{3,7\}$, $\{4,6\},\{5\}$. But there are 6 classes and 7 integers.

16. Let a_i = numbers of hours worked from first day through i-th day, $i = 1,2,\ldots,49$. Then $1 \leq a_i < a_{i+1} \leq 77$, and hence $21 \leq a_i + 20 < a_{i+1} + 20 \leq 97$.

The set of 98 numbers $\{a_i\} \cup \{a_i + 20\}$ range between 1 and 97. So for some $i < j$, $a_j = a_i + 20$, i.e. from day $i+1$ through day j, 20 hours worked.

18. See Exercise 31 in Supplement 2 of Chapter I

Appendix 5

1. R Bu G Y
2. W R Bu R or Y W G R
3. Y Bu Bu W
4. G W G Bu or G Bk G Y
5. fourth guess= Bu Bk Bk Bu or Bk Bu Bu Bk. There are 8 possibilities after the first 3 guesses to be distinguished: G G Bk W, G Bk Bk W, G Y G G , G Bu Bk Bk, G Bu Bu R, G Bu R R, G Bu Bk G, G Bu Bk Bu.
6. Best fourth guess= Bk Bu Bu Bu or Bu Bu Bu Bk (many other possible guesses).   There are 3 possiblilites after the first 3 guesses to be distinguished: Bk Bu Bu Bu, Bu Bu Bu Bk, W Y W W.
7. G Y R Bk
8. Bu Bu Y O or Bu O W Bk
9. Bu Bu Bk R, Bu Bk R Bu, Bk Bk R Bu, Bu Bu Bk R, W Bu R Bu, W Bu R W, W W R Bu.
10. W O G O Bu
11. O R Y Bu P
12. 1/1296
15. 9
17. Use all 4 pegs same color. Then 5^4 possibilities remain, or 6^4-5^4 = 671 eliminated.